BASIC STATISTICS

T. J. HANNAGAN

M
MACMILLAN EDUCATION

© T. J. Hannagan 1982

All rights reserved. No part of this
publication may be reproduced or
transmitted, in any form or by any
means, without permission.

First published 1982

Published by
MACMILLAN EDUCATION LIMITED
Houndmills Basingstoke Hampshire
RG21 2XS and London
Associated companies in Delhi
Dublin Hong Kong Johannesburg
Lagos Melbourne New York
Singapore and Tokyo.

Printed in Hong Kong

This book is also available under the
title *Mastering Statistics* published
by Macmillan Press.

HUMBERSIDE LIBRARIES

Div. C N/o ALLOC.

Class 519.5

ISBN 0 333 32687 3

ALS No. H1673386

Plessey No. B26 169 080 X

To Yvonne, Jane, Kathleen, Liam, Kevin

CONTENTS

Preface xi
Acknowledgements xiii

1 **Information**
  1.1 Forms of information 1
  1.2 Definition of statistics 1
  1.3 Statistical information 2
  1.4 Numerical data 2
  1.5 The abuse of statistics 4
  1.6 The use of statistics 5
  1.7 Primary and secondary data 6
  1.8 Secondary statistics 8

2 **Sources of information**
  2.1 Where to find information 10
  2.2 Statistical information in business 10
  2.3 Government statistics 11
  2.4 Government statistics in the UK 12
  2.5 The standard industrial classification 14
  2.6 The use of libraries 15
  2.7 Problems with official statistics 15
  2.8 How to profit from official statistics 16
  2.9 Conclusions 17

3 **The accuracy of information**
  3.1 Approximations 19
  3.2 Degrees of tolerance 20
  3.3 Error 21
  3.4 Rounding 22
  3.5 Absolute error 24
  3.6 Relative error 25
  3.7 Biased, cumulative or systematic error 25
  3.8 Unbiased or compensating error 26
  3.9 Calculations involving approximation and error 26
  3.10 Spurious accuracy 30

CONTENTS

4	**Collecting information**		4.1	Surveys	33
			4.2	Survey methods	34
			4.3	Observation	35
			4.4	Interviewing	37
			4.5	Questionnaires	40
			4.6	Collecting information	41
5	**How to sample**		5.1	Why sample?	43
			5.2	Advantages of sampling	43
			5.3	Objectives of sampling	44
			5.4	The basis of sampling	45
			5.5	Sampling errors	46
			5.6	Sample size	47
			5.7	Sample design	47
			5.8	Bias in sampling	48
			5.9	Simple random sampling	49
			5.10	Systematic sampling	50
			5.11	Random route sampling	51
			5.12	Stratified random sampling	51
			5.13	Quota sampling	52
			5.14	Cluster sampling	54
			5.15	Multi-stage sampling	55
			5.16	Multi-phase sampling	57
			5.17	Replicating or inter-penetrating sampling	57
			5.18	Master samples	58
			5.19	Panels	58
6	**How to use figures**		6.1	The need to use figures	60
			6.2	The vocabulary of mathematics	60
			6.3	Basic arithmetic	61
			6.4	The sequence of operations	62
			6.5	Simple arithmetic	63
			6.6	Fractions	63
			6.7	Decimals	65
			6.8	Percentages	67
			6.9	Powers and roots	68
			6.10	Ratios	69
			6.11	Proportions	69
			6.12	Elementary algebra	70
			6.13	Levels of measurement	70
			6.14	Financial mathematics	72
			6.15	Aids to calculation	75
			6.16	Symbols of mathematics	80

7	**How to present figures (1)**	7.1 The aims of presentation	82
		7.2 Tabulation	83
		7.3 Classification	85
		7.4 Frequency distributions	87
		7.5 Reports	89
		7.6 Histograms	90
		7.7 Frequency polygons	94
		7.8 Frequency curves	96
8	**How to present figures (2)**	8.1 Bar charts	99
		8.2 Pie charts	104
		8.3 Comparative pie charts	106
		8.4 Pictograms	107
		8.5 Comparative pictograms	109
		8.6 Cartograms or map charts	109
		8.7 Strata charts	110
		8.8 Graphs	110
		8.9 Semi-log graphs	116
		8.10 Straight-line graphs	118
		8.11 The gantt chart	119
		8.12 Break-even charts	120
		8.13 Z-charts	122
		8.14 Lorenz curves	123
		8.15 Presentation and perception	125
9	**Summarising data: averages**	9.1 The role of the average	129
		9.2 The arithmetic mean	130
		9.3 The arithmetic mean of a frequency distribution	132
		9.4 The arithmetic mean of a grouped frequency distribution	134
		9.5 Advantages and disadvantages of the arithmetic mean	137
		9.6 The median	137
		9.7 The quartiles	141
		9.8 The mode	144
		9.9 The geometric mean	148
		9.10 The harmonic mean	148
10	**Summarising data: dispersion**	10.1 Dispersion	151
		10.2 The range	155
		10.3 The interquartile range	156
		10.4 The standard deviation	159
		10.5 The variance	169
		10.6 The coefficient of variation	170
		10.7 Conclusions	170

CONTENTS

11	**Statistical decisions**	11.1	Estimation		173
		11.2	Probability		173
		11.3	The normal curve		175
		11.4	Probability and the normal curve		178
		11.5	Standard error		179
		11.6	Tests of significance		181
		11.7	Confidence limits		183
		11.8	Statistical quality control		184
		11.9	Conclusion		186
12	**Comparing statistics: index numbers**	12.1	In general		188
		12.2	Calculation of a weighted index number		190
		12.3	Chain-based index number		192
		12.4	Problems in index-number construction		193
		12.5	The Index of Retail Prices		194
13	**Comparing statistics: correlation**	13.1	What is correlation?		198
		13.2	Scatter diagrams		200
		13.3	Correlation tables		203
		13.4	The product moment coefficient of correlation		204
		13.5	The meaning of r		206
		13.6	Coefficient of rank correlation		206
		13.7	Spurious correlation		208
		13.8	Regression		209
14	**Trends and forecasting**	14.1	Trends and forecasting		211
		14.2	Time series		212
		14.3	Moving averages		213
		14.4	Seasonal variations		215
		14.5	Irregular and residual fluctuations		219
		14.6	Linear trends		220
		14.7	Conclusions		226
		14.8	The next step		227

Appendixes

A.1	*Reading list*	229
A.2	*Area table*	230
A.3	*Square-root tables*	232
A.4	*Logarithm and antilogarithm tables*	236
	Index	241

PREFACE

The aims of this book are to provide a sound statistical foundation and appreciation for a wide range of readers, some of whom may have very limited mathematical knowledge. The book includes a chapter on simple arithmetic as a reminder of basic mathematical techniques.

The book is designed to provide a comprehensive appreciation of statistical sources, concepts, methods and applications, and therefore it provides a firm basis for further quantitative and numerical study.

It aims at stressing the relevance of statistics to everyday life and in a wide range of business situations. Therefore the book is designed for:

(a) the *student* who is studying for a range of examinations, courses and modules;
(b) the *general reader* who wants to have a greater understanding of events, facts and figures in everyday life;
(c) the *businessman* who needs to understand, interpret and make use of statistical information and to use statistical techniques in industry and commerce.

This book provides a comprehensive coverage of the statistics required in most business studies, professional and management courses, as well as Social Science and Public Administration Degree courses.

At the end of each chapter assignments are included to reinforce the information in the chapter. These assignments are designed to provide for both individual learning and for group activity. While it is quite possible to read and understand the book without completing any of these assignments, they are aimed at illustrating the practical uses of a particular technique or concept as well as providing practice for examinations.

On completion of this book the reader should be aware of the main statistical sources of information, understand numerical concepts and be proficient in basic statistical methods. Also the reader should be able to apply a variety of statistical techniques to help solve business and everyday problems.

ACKNOWLEDGEMENTS

The author and publishers wish to thank the Controller of Her Majesty's Stationery Office for permission to use tables from *Social Trends 10* and *Economic Trends* (April 1980) and The Macmillan Education Ltd for permission to reproduce square-roots, logarithms and antilogarithms tables from F. Castle, *Four-Figure Mathematics Tables* (Macmillan, 1960).

CHAPTER 1

INFORMATION

1.1 FORMS OF INFORMATION

People receive large quantities of information every day through conversations, television, the radio, newspapers, posters, notices and instructions. Motorists approaching major road junctions have so much information to absorb in the form of road signs, direction indicators and the flow of traffic using the junction that accidents are caused because they are unable to see, understand and act on the information sufficiently rapidly.

It is just because there is so much information available that people need to be able to absorb, select and reject it, to sift through facts to pick out what is interesting and useful. In everyday life as well as in business and industry, certain information is necessary and it is important to know where to find it or how to collect it.

As consumers, everybody has to compare prices and quality before making a decision about what goods to buy. As employees, people want to compare wages and conditions of work. As employers, firms want to control costs and expand profits. Therefore everybody collects, interprets and uses information, much of it in a numerical or statistical form.

1.2 DEFINITION OF STATISTICS

Statistics is concerned with scientific methods for collecting, organising, summarising, presenting and analysing data, as well as drawing valid conclusions and making reasonable decisions on the basis of this analysis.

Data (singular 'datum') are things known or assumed as a basis for inference. Statistics is concerned with the systematic collection of numerical data and its interpretation.

The word 'statistics' is used to refer to:

(i) numerical facts, such as the number of people living in a particular town;

(ii) the study of ways of collecting and interpreting these facts.

It can be argued that figures are not facts in themselves; it is only when they are interpreted that they become relevant to discussions and decisions.

1.3 STATISTICAL INFORMATION

The main function of statistics is to provide information which will help in making decisions. It can be argued that to make decisions without the help of statistics is asking for trouble; rather like asking a doctor to diagnose a patient's illness without the doctor being able to collect information on the patient's condition, such as temperature, pulse, blood pressure and so on. Statistics provides this type of information by providing a description of the present, a profile of the past and an estimate of the future.

The ease of processing numerical data has increased its use and availability; however, there is a tendency to avoid numerical information in favour of written, visual or verbal information. Few newspapers publish any volume of statistics and the media in general tends to ignore the range and variety of statistical publications available (see Chapter 2). There is a tendency to translate statistics into a verbal form. The headline 'High Unemployment but Consumer Boom Continues' is a statistical statement. It is an interpretation of data on the levels of employment and of data on sales figures. Words like 'greater', 'more', 'higher', 'lower', 'small', 'insignificant', all imply quantification. To say or write that unemployment is high may be a summary of a complex table of figures.

One of the interesting exceptions to this avoidance of statistics is in sport, where complicated tables and figures are accepted as normal. People have little difficulty in understanding football league tables, batting averages or pools coupons. Perhaps this indicates that everybody is capable of understanding complicated numerical information if they are sufficiently well motivated.

If Table 1.1 can be understood and addition, subtraction, multiplication and division present few difficulties, then mastering statistics will present few problems. Some basic mathematical rules are outlined in Chapter 6.

1.4 NUMERICAL DATA

It is interesting to notice that to be illiterate is considered unacceptable, but to be innumerate is not. However, many facts can be given sensibly only in numerical form. This includes the fairly complicated data provided in the football league table (Table 1.1). Also it includes most facts about money.

Table 1.1   Football League table

Football Club	P	W	D	L	Goals F	Goals A	Pts
Liverpool	38	23	9	6	75	28	55
Man. Utd	39	22	10	7	61	31	54
Ipswich	40	21	9	10	66	37	51
Arsenal	37	16	14	7	48	29	46
A. Villa	38	14	14	10	46	43	42
Southampton	39	16	9	14	56	48	41
Wolves	37	17	7	13	49	41	41
Nottm F.	36	17	6	13	55	40	40
West Brom.	39	11	17	11	53	48	39
Middlesbrough	37	14	11	12	42	37	39
Crystal P.	40	12	15	13	41	46	39
Coventry	39	16	7	16	54	61	39
Leeds	40	12	14	14	43	47	38
Tottenham	39	15	8	16	50	59	38
Norwich	39	11	14	14	51	60	36
Brighton	40	11	14	15	47	56	36
Man. City	40	11	13	16	40	62	35
Stoke	39	11	10	18	42	56	32
Everton	38	8	15	15	41	50	31
Derby	40	10	8	22	42	62	28
Bristol C.	38	8	12	18	30	57	28
Bolton	40	5	14	21	38	72	24

Everybody is affected by inflation as consumers, wage earners and employers. With a few exceptions, which tend to highlight the general rule, prices of goods and services rise week by week and year by year. A house priced at £20,000 today might have been priced at £2000 twenty years ago and might have a price of £120,000 in twenty years' time. This type of increase is well known, but confusion is caused by the various statements made about inflation and the methods used to measure it. On the same day it may be stated that:

(a) 'inflation is 6%',
(b) 'inflation is running at an annual level of 10%',
(c) 'inflation this year is running at 15%',
(d) 'the average level of inflation is 18%'.

All of these statements could be correct and all at the same time. The differences between them arise because they are statements about different aspects of inflation, they represent different ways of calculating the level of inflation. The statements are incomplete, because they do not make it clear to what aspect of inflation the percentages relate.

(a) **'Inflation is 6%'** could be arrived at by multiplying the rise in prices during one month by twelve to arrive at an annual figure: $\frac{1}{2}$% monthly price rice multiplied by 12 = 6%.

(b) **'Inflation is running at an annual level of 10%'** could represent the rise in the general price level in the last twelve months. Prices in general might have risen by $\frac{1}{2}$% in the last month, but may have increased faster in the previous eleven months to produce an annual level of 10%: monthly inflation might be 25%, 18%, 16%, 15%, 12%, 11%, 8%, $5\frac{1}{2}$%, 5%, 3%, 1%, $\frac{1}{2}$%, making a total of 120, which divided by 12 = 10%.

(c) **'Inflation this year is running at 15%'** could refer to just this particular year, that is since January 1 (or it might refer to the financial year). This would mean that at the end of July 15% inflation would represent average price rises over a period of seven months: 25%, 18%, 16%, 15%, 12%, 11%, 8%, making a total of 105, which divided by 7 = 15%.

(d) **'The average level of inflation is 18%'** could be looking at inflation over the last three years: say annual rates of inflation of 24%, 20%, 10%, which makes a total of 54%, which divided by 3 = 18%.

This illustrates the point that numerical data by itself does not provide any information at all. A set of figures, such as 8%, $5\frac{1}{2}$%, 5%, 3%, 1%, $\frac{1}{2}$%, are meaningless unless they are related in some way. Directly it is known that these figures refer to the percentage inflation rates over the last six months, then the figures have meaning. They still need to be interpreted. For example, from a knowledge of the meaning of inflation it could be said that these figures indicate a falling monthly rate of inflation.

Interpretation is a matter of judgement based on knowledge, for example the knowledge of what is meant by the term 'inflation'. It is at least as important that a figure or the result of a calculation is understood as it is that it is accurate. There is little point in arriving at a correct answer to a calculation if it is not known what it means.

1.5 THE ABUSE OF STATISTICS

'You might prove anything by figures'. (CARLYLE)
'There are three kinds of lies: lies, damn lies, and statistics.' (DISRAELI)
'Don't be a novelist - be a statistician, much more scope for the imagination'. (MEL CALMAN)
'He uses statistics as a drunken man uses a lamp post - for support rather than illumination. (ANDREW LANG)

These statements are both right and wrong, it is all a matter of interpretation. If the objective is to show that the rate of inflation is low, it would be possible to say that is it running at $\frac{1}{2}$% (statement (a) in Section 1.4), on the basis that prices rose by this amount last month. If the objective is to prove that the level of inflation is high, it would be possible to take the highest available measure (see statement (d) in Section 1.4) and state that inflation is running at 18%, on the basis that the average level of price rises over the last three years has been 18%.

In fact, it would be accurate to say that 'you can prove nothing with statistics'. Figures are used to support theories, opinions and prejudices. It can be argued that in the above examples it was decided first of all that the objective was to show that the rate of inflation was low or high and then statistics were found to support these decisions.

It can be argued that statistics are used to support a case, either to make it appear 'scientific', or because by collecting and examining all the statistical evidence relevant to a case it will in fact increase its validity.

Ideally, the collection and analysis of statistics is carried out as accurately and objectively as possible. In practice, statistics are at times employed to sensationalise, inflate, confuse or oversimplify. Therefore it is important to be realistic about the way in which statistics are collected and summarised.

1.6 THE USE OF STATISTICS

It can be argued that the principal function of statistics is to narrow the area of disagreement which would otherwise exist in a discussion and in that way help in decision making. Statistics can be a stabilising force, dispelling rumour and uncertainty, helping to solve arguments arising from individual cases or circumstances by providing a factual foundation to debates and decisions which would otherwise be dominated by subjectively based theories and opinions.

In a sense statistics are the opposite of the 'Grandmother theory'. This theory runs on the lines of 'my Grandmother (or Grandfather) smoked fifty cigarettes (drank a bottle of whisky/never did any exercise/ate like a horse) every day and lived to be one hundred and five years old; therefore smoking (drinking, lying around, over-eating) cannot be bad for you'. Statistics do not prove that everybody who smokes heavily dies young, but the statistical evidence does show that people are more likely to die young if they smoke heavily than if they do not.

It is this feature of systematic collection of data that distinguishes statistics from other kinds of information. Statistics provide a method of systematically summarising aspects of the complexities of social and economic problems.

It is useful to divide statistics into:

(a) **Descriptive statistics**: including the presentation of data in tables and diagrams as well as calculating percentages, averages, measures of dispersion and correlation, in order to display the salient features of the data and to reduce it to manageable proportions.

(b) **Inductive statistics**: involving methods of inferring properties of a population on the basis of known sample results. These methods are based directly on probability theory.

In fact most knowledge is 'probability knowledge' in the sense that it is possible to be absolutely certain that a statement is true only about statements of a restricted kind. Statements such as 'I was born after my father', 'a black cat is black', 'one plus one equals two' are tautologies or disguised definitions, and although they are true they are of limited value in providing information. Most information is based on sample data rather than on a complete survey of a population, and probability theory is the basis of the areas of statistics that are not purely descriptive (in particular see Chapters 5 and 11).

The emphasis in the work of the statistician has shifted from a backward-looking process to current affairs and to proposed future operations and their consequences. Every business organisation and government department uses statistics in its daily work. There has been an enormous development in the collection, processing and storing of statistical information in recent years, which has been assisted by the growth of computer technology. H. G. Wells suggested that 'statistical thinking will one day be as necessary for efficient citizenship as the ability to read and write'. Perhaps that day has come.

Many people now have pocket calculators and if they do not need them for business or for study they may still use them as consumers or taxpayers or sports enthusiasts. Professional tennis players' ratings and golfers' handicaps are subject to computer programmes. The use of electronic means of processing data has removed much of the repetitive drudgery of statistical calculations, but has made it more than ever necessary to understand statistical material. Facts and figures cannot be accepted on their 'face value'; it is necessary to know something about the methods used to collect the data and to summarise it in order to understand its imperfections and strengths.

1.7 PRIMARY AND SECONDARY DATA

Numerical data can be divided into primary and secondary data:

(a) **Primary data**: this is collected by or on behalf of the person or people who are going to make use of the data. This collection of data involves all the survey and sample methods described in Chapters 4 and 5. Once data has been collected, processed and published it becomes secondary data.

(b) **Secondary data**: this is used by a person or people other than the people by whom or for whom the data was collected. It is 'second-hand' data. Few people collect their own statistical material except on a small scale and almost everybody collects some statistical material on a small scale, although not necessarily very systematically. Consumers compare prices at a number of shops before making a purchase; firms collect data about aspects of their business, such as output and sales. However, for many general purposes and for collections of large-scale statistical material and for material collected systematically, most people use tabulations produced by others.

Because this is second-hand data, it is important to know as much about it as possible just as it is important to know as much as possible about a second-hand car before buying it. It is useful to know how the data has been collected and processed in order to appreciate the reliability and full meaning of the statistics. Tables can be misleading: for example a table may show that a firm has increased its output of a small component this year as compared with last year by 100%. This result appears less impressive if it is known that the firm produced only 5 of these components last year and 10 this year. If these numbers had been 5000 and 10,000, the result would have been more impressive and the percentage more appropriate.

The main points to be considered in the use of secondary data are:

(i) How the data has been collected.
(ii) How the data has been processed.
(iii) The accuracy of the data.
(iv) How far the data has been summarised.
(v) How comparable the data is with other tabulations.
(vi) How to interpret the data, especially when figures collected for one purpose are used for another.

Generally with secondary data, people have to compromise between what they want and what they are able to find. However, there are great advantages in using secondary data:

(i) it is cheap to obtain - many government publications are relatively cheap and libraries stock quantities of data produced by the government, by companies and other organisations;
(ii) this means that large quantities of secondary data is easily available - although it may not be exactly what is wanted;
(iii) there is a great variety of data on a wide range of subjects;
(iv) much of the secondary data available has been collected for many years and therefore it can be used to plot trends.

Secondary data is of value to:

(i) *The government* – to help in making decisions and planning future policy;
(ii) *Business and industry* – in areas such as marketing and sales, in order to appreciate the general economic and social conditions, and to provide information on competitors;
(iii) *Research organisations* – by providing social, economic and industrial information.

Published statistics tend to be either the product of an 'administrative process' or to come from especially conducted enquiries or surveys. Statistics on unemployment are a good example of the former. They are collected mainly as part of the process of registration by individuals as unemployed in order to be eligible for unemployment benefit. Therefore the statistics are collected as part of an administrative process and are not entirely appropriate for other uses.

In fact these figures are used to estimate the number of people unemployed, but they are not an accurate record because large numbers of people who are looking for work do not register as unemployed. This includes people who are not eligible for benefit, such as housewives, and people who expect to find work at any moment and do not bother to register. Also, the unemployment statistics provide no information on underemployment; that is people who are at work but are not working at full capacity. Therefore at best official unemployment statistics provide a considerable underestimation of the level of unemployment.

Statistics from opinion polls, the Census of Population and the Family Expenditure Survey are good examples of surveys and inquiries designed solely to obtain information (for more information on these see Chapter 2). The Census of Population in the UK is carried out every ten years in an attempt to discover a wide range of information on households. This information may help to formulate administrative decisions but it is not collected as part of an administrative process.

Surveys and enquiries can be more easily used as secondary data than the bi-product statistics produced as part of an administrative process. Not only has the data been collected for the purpose of providing information but also details are available describing how the survey or census has been conducted.

1.8 SECONDARY STATISTICS

Statistics compiled from secondary data are termed *secondary statistics*. Therefore calculations made on the basis of government figures on, for example, unemployment are described as secondary statistics. The tables of unemployment figures on which the calculations are based are secondary data.

The problems of using secondary data apply equally to secondary statistics. Anyone using published statistics needs to consider the purpose for which they were originally compiled. Secondary statistics are common because secondary data is so easily available.

There are vast amounts of statistical information available and it is important to know to what uses this data can be put and how to make use of it. Before it can be used it has to be found and therefore it is very important to know where to find it.

ASSIGNMENTS

1 Write a commentary in about 750 words on the Football League table (Table 1.1) in Section 1.3, to bring out as much information as possible.

2 Discuss (in writing or verbally in a group) what is meant by the term 'information'. Study the example of inflation used in Section 1.4. Find a similar example and describe how this can illustrate the problem of the interpretation of data.

3 What are the main problems involved with using secondary data? Illustrate these problems by reference to any table of secondary data.

4 Consider the kind of problems a person will face who cannot read or write. How far are these problems similar to those of a person who cannot count or understand figures?

5 How far is it possible to agree with the statements made at the beginning of Section 1.5? Find examples to support these statements.

CHAPTER 2
SOURCES OF INFORMATION

2.1 WHERE TO FIND INFORMATION

Government departments, employers' federations, trade associations, trade unions, private firms, professional institutes, public and private research organisations collect and publish large quantities of statistical information.

The problem is to find the information required, or the closest approximation to it, when it is required. In order to do this it is useful to have some idea of the types of information available and where to find it.

Statistical information can be divided into:

(a) **Micro-statistical information**: data produced by private firms and private organisations. It is information specific to the business organisation, compiled in the process of monitoring activities of the business unit. It may include information about aspects of production, marketing, labour supply and so on.

(b) **Macro-statistical information**: data produced by the public sector and which relates to the country as a whole rather than one organisation or unit. This data includes information about population, finance, education, unemployment and so on.

2.2 STATISTICAL INFORMATION IN BUSINESS

Business organisations both collect and use statistical information in the process of carrying out their activities. This collection and use of statistical information depends on the size and complexity of the business. For example, the local butcher will have different statistical needs to those of a major insurance company.

In general, businesses may be interested in:

(a) Changes in the national economy - including factors such as interest rates and levels of regional unemployment.

(b) Statistics relating to the particular industry in which the firm operates — particularly information in such areas as wage rates, prices, levels of output which enables the firm to make comparisons between its performance and that of its competitors.

Businesses compile quantities of data in connection with monitoring their activities. This data may be in such areas as:

(a) **Production**: a firm may collect data as a result of using quality control, and checking on various aspects of its work including labour turnover, production progress, stocks and inventories, wage rates, accident rates, sickness rates, absenteeism rates, unit costs, raw material costs.

(b) **Marketing**: firms collect data from market research, advertising, sales results, distribution costs and so on.

(c) **Administration**: business administration involves auditing, information about loans and borrowing, government requirements such as taxation and so on, as well as the overall organisation of the business. In any business the 'office work' is concerned with compiling and processing information and storing and retrieving it. In the butcher's shop this may be carried out by one or two people by hand in a small office or at the dining-room table; in the large insurance company there will be specialist departments, employing large numbers of people, and using sophisticated equipment.

(d) **Personnel**: where information on recruitment and training are produced and also on industrial relations and wage rates.

Businesses use statisticians to provide clear and concise summaries of the firm's business activity, to carry out surveys and samples to produce trends and forecasts, which can provide the basis for decisions. The statistical practices in firms vary considerably depending on the size of the firm and the type of business. There are no standard forms or methods used by every firm, however the statistics covered in this book are standard and can be applied to any business.

2.3 GOVERNMENT STATISTICS

Governments produce statistics because:

(i) they want to be able to measure the effects of their policies;
(ii) they want to monitor the effects of external factors on their policies;
(iii) they need to be able to assess trends so that they can plan future policies.

Much of the information collected by governments is widely available. The British Government has produced the *Annual Abstract of Statistics* for over a hundred years and since its inception the range and detail of the government statistics produced has greatly increased. This is a reflection of the greater needs and complexity of society and the growth of the economy.

Governments rely on private firms, institutions and individuals to provide the basic data. Organisations of all kinds and every adult person has government forms to complete. For individuals these forms include income tax returns, population census forms and the electoral register form. Businesses receive a range of forms to complete, concerned with taxation, prices, costs, output and wage rates. All these forms provide the raw data which is then processed, analysed and collated to form the basis of the tables, graphs, averages and indexes produced by the government and its departments.

In their turn business firms use these collations of information for their corporate planning. Businesses need a continuous assessment of their position and, as well as collecting their own statistics to monitor this, they also need to be able to assess their position with reference to the business and economic environment in which they operate.

2.4 GOVERNMENT STATISTICS IN THE UK

The United Kingdom Government Statistical Service comprises the statistical divisions of all the major government departments and the Central Statistical Office (CSO) which co-ordinates the system. Also there are two major collecting agencies, the Business Statistics Office and the Office of Population Censuses and Surveys (OPCS) which includes the registration division of the Registrar General's Office.

The Government Statistical Service exists to serve the needs of the government, but much of the information compiled is usable by business management and parts of the system have been especially designed so that this is possible.

The information is supplied to the press so that it is available quickly, and in more detail it is published regularly in a range of digests and monitors. The CSO provides a clearing house for enquiries and contributes to the Post Office Prestel television data information service. All this information can be bought at Her Majesty's Stationery Offices (HMSO), which act as the government's bookshop. Also, all of it is available at major libraries and much of it at local libraries.

The CSO has produced a *Guide to Official Statistics*, which aims to cover all official and some important non-official sources of statistics for the United Kingdom. It contains broad content descriptions of all relevant publications containing a significant amount of statistical information. It does not attempt to list every commodity, service, occupation and other area or subject for which statistics exist, but it does attempt to provide a good indication of whether or not a particular subject is included in a given source. It covers international publications only where they provide a unique source of UK data.

The subjects covered by this *Guide* give an indication of the material available. These subjects include:

Climate	Distribution
Population	Public Services
Vital Statistics	Prices
Social Statistics	National Income and Expenditure
Labour	
Agriculture	Public Finance
Production Industries	Financial and Business Transactions
Transport	
	Overseas Transactions

Also general and non-government sources of statistics are included, such as:

Whitaker's Almanack – which provides basic statistical information in a compact form.

Statesman's Year-Book – which provides international statistics.

Municipal Year Book – which provides basic data for local authority areas.

Guinness Book of Records – which provides numerical data not readily available elsewhere.

To supplement the *Guide*, the CSO produces a booklet: *Government Statistics: a brief guide to sources*, which provides a summary of the larger *Guide* and is brought up to date each year.

The CSO publishes a quarterly, *Statistical News*, which provides a comprehensive account of current developments in British official statistics. Also the CSO has produced a booklet: *Profit from Facts*, which illustrates the use firms can make of government statistics; and a booklet called *Facts from Your Figures*, which is dedicated to all 'those conscientious people who fill in statistical forms but sometimes wonder why ... '. It illustrates why official statistics are needed and how they are collected.

Perhaps the most readily available and frequently used government statistical publications are:

(a) *Monthly Digest of Statistics* – which is a collection of the main statistical series from all government departments;

(b) *Financial Statistics* – which is a monthly summary of key financial and monetary statistics;

(c) *Economic Trends* – which is a monthly commentary and selection of tables and charts providing a background to trends in the UK economy;

(d) *British Business* – (formerly *Trade and Industry*), this is a weekly compilation of statistics and commentary from the Departments of Industry and Trade;

(e) *Employment Gazette* - which is a monthly publication from the Department of Employment including articles, tables and charts on manpower, employment, unemployment, hours worked, wage rates, retail prices and so on;

(f) *Economic Progress Report* - this is a monthly pamphlet published by the Treasury's Information Division and contains articles on economic subjects, on government economic policy and the Treasury's assessment of the economic situation, along with tables and charts of economic indicators.

These are all monthly or weekly publications and all of them are mainly concerned with statistics in economic and business areas. There are a number of annual publications which cover these and other areas:

(g) *Annual Abstract of Statistics* - which contains more series than the Monthly Digest;

(h) *Social Trends* - which brings together key social and demographic series;

(i) *Regional Statistics* - which provides a selection of the main regional statistics;

(j) *National Income and Expenditure 'Blue Book'* - which gives detailed estimates of the national accounts;

(k) *Family Expenditure Survey* - which shows in detail household income and expenditure.

As well as these publications there is a range of detailed booklets and volumes on specific areas, such as transport and education. Also the government produces a range of Indexes (or Indices) including those on Retail Prices, Wholesale Prices, Industrial Production, Exports and Imports and Retail Sales. Census results are produced in a series of volumes, most notably for the Census of Population, the Census of Production and the Census of Distribution.

This summary of the official sources of data in the UK gives an indication of how well covered are areas such as business, the economy and finance. These sources are further supported by publications such as the *Bank of England Quarterly Bulletin*, the *Reviews* produced by the major banks and the journals produced by a range of professional institutions (such as the British Institute of Management's *Management Today*).

2.5 THE STANDARD INDUSTRIAL CLASSIFICATION

Statistics can be compared only when they are collected on a similar basis. Official statistics in the UK are collected from a wide variety of sources and the Standard Industrial Classification (SIC) has been drawn up to ensure that statistics issued by different departments are comparable.

It is a system of classification of establishments according to industry. This includes all economic activities such as agriculture and public administration as well as production industries. It has been compiled to conform with the organisation and structure of industry as it exists within the UK.

The great advantage of the SIC is that reference can be made to any official statistics with the knowledge that there will be uniformity and comparability in the classifications used.

2.6 THE USE OF LIBRARIES

The problem with finding official statistics and private sources in libraries is that statistical publications are rarely the responsibility of individual authors. They are the product of a government department or agency, or produced by a private firm or institution. In library catalogues the term 'author' is usually interpreted as the responsible body.

For example: the author of the Census of Population volumes in England and Wales is listed as the 'Office of Population Censuses and Surveys'. The author of the *Monthly Digest of Statistics* is listed as the 'Central Statistical Office'.

Journals are not listed under author but under title. Many statistical publications in the UK are catalogued under the broad headings of Great Britain or Northern Ireland.

For example: Great Britain, Department of Education, Committee on Higher Education, *Higher Education* (Robbins Report).

When sources of statistical information have been located it is possible to collect the required data. In the process of this collection it is useful to:

(i) Make sure that the best material has been located for the particular purposes of the enquiry.
(ii) In order to save time, be clear about exactly what is required for the enquiry and extract only those statistics that are necessary.
(iii) Where there are gaps in the required data, try to find material to fill these gaps.

2.7 PROBLEMS WITH OFFICIAL STATISTICS

It was pointed out in Section 1.7 that there are problems involved in using secondary data. There often has to be a compromise between the statistics that are wanted and what is available. Official statistics are collected to cover activities which appear to be relevant to the administrative needs of the time. They are a by-product of administrative processes and the figures collected for one purpose may not be suitable for another.

These problems have been increasingly recognised in the UK by the Central Statistical Office. In many cases official statistics are accompanied by phamphlets or volumes which explain how the data has been collected and processed, and the levels of accuracy and reliability which can be expected. Also there are attempts to indicate how these sources can be useful to business, for example by the publication of booklets such as *Profit from Facts*.

2.8 HOW TO PROFIT FROM OFFICIAL STATISTICS

Government statistics can help firms in a variety of areas:

(a) In **marketing**, official sources can help a firm to:

(i) assess the market share trends in a number of product fields;
(ii) watch the size and growth of existing and potential markets;
(iii) count the number of potential customers, including regional patterns
(iv) see how people spend their money;
(v) check on distribution channels;
(vi) check on price changes;
(vii) use retail sales and stock movements to assist short-term sales forecasting;
(viii) assess the possibility of meeting foreign competition in home markets;
(ix) estimate world markets;
(x) fix quotas for area salesmen through regional statistics;
(xi) estimate the effect of weather on business.

(b) In **buying**, official statistics can help a firm to:

(i) be aware of the sales trends of materials and goods;
(ii) trace the price movements of materials.

(c) In **personnel**, official statistics help a firm to watch trends by industry and by region in unemployment, vacancies, earnings, overtime, wage rates, hours of work, industrial disputes and compare these trends with the firm's own situation.

(d) In **financial control**, official statistics enable firms to compare their results with those of other firms. These results include such areas as:

(i) operating ratios, such as net output per head, stocks as a percentage of sales, wages per £ of total sales, turnover per employee;
(ii) labour costs;

(iii) company finance, including aggregated balance sheets, appropriation accounts, sources and uses of funds, income, interest and dividend payments as a percentage of assets;
(iv) current replacement costs of fixed assets and stocks can be assessed.

2.9 CONCLUSIONS

There is a mass of secondary sources of data both in the form of official statistics and in the form of private publications. In the process of monitoring their business activities, firms accumulate statistics which are mainly for their own benefit but also are used as a base for official sources through the various returns firms make to government departments. Firms can compare their own statistics with those available for their industry and for the economy.

Much of the statistical information available can be obtained relatively cheaply and easily. Exactly how it is used will depend on the size and nature of a particular organisation and the importance attached by management to monitoring progress and to planning and development.

ASSIGNMENTS

1 Outline and discuss the type of statistical information gathered by any one organisation in the normal process of its work. Consider how this information could be used to help the organisation in the future.

2 Find the latest unemployment figures for one region and for one town in that region. Compare these figures with the national level of unemployment. Are the trends of unemployment in the region and the town the same as those for the whole country?

3 Consider the uses of official statistics to a particular firm.

4 Discuss the information contained in Table 2.1. What are the problems of using the information from a table of this kind?

5 Find television viewing figures and write a report on the following questions:
(i) How many hours of television does the average child and the average adult view each week?
(ii) Do children view more television than adults, and men more than women?
(iii) What influence does unemployment and retirement have on average weekly viewing hours?

Table 2.1 direct and indirect taxes as percentages of gross domestic product: international comparison[1]

	Direct and indirect taxes[2] as percentages of GDP			Direct taxes[3] as percentages of GDP		
	1966	1971	1976	1966	1971	1976
Denmark	27.6	39.8	39.4	15.1	24.6	24.4
Finland	25.3	29.0	34.1	11.8	14.8	20.8
Norway	27.4	32.1	34.1	13.6	14.6	16.9
Sweden	29.6	32.1	33.6	18.0	19.7	21.9
Netherlands	26.0	28.9	31.9	16.2	18.3	21.3
Luxembourg	19.4	20.5	28.6	12.0	13.4	18.9
Ireland	19.7	23.8	28.2	5.9	8.8	11.4
Belgium	21.9	24.0	28.0	9.9	12.3	16.9
Austria	24.4	25.5	26.5	11.5	12.2	13.3
Germany	22.3	23.7	25.6	12.6	13.9	16.7
United Kingdom	21.5	23.5	25.5	12.3	14.1	16.8
New Zealand	17.5	19.9	24.1	10.8	13.0	17.2
Switzerland	15.6	16.8	21.6	9.8	11.1	16.0
France	18.9	19.5	21.0	5.9	6.5	8.5
Australia	15.7	17.2	20.9	8.5	10.2	13.4
Italy			18.7			9.1
USA	14.9	17.3	17.2	10.2	12.2	12.5
Greece	15.3	16.6	16.2[4]	4.8	5.9	5.4[4]
Portugal	9.8	12.4	15.5	2.5	3.5	5.4
Japan	9.7	10.1	10.5	5.2	6.1	7.1
Spain	9.8	9.3	10.0	3.1	3.6	5.0

[1] The countries shown are those for which data are available in the source publication, and are ranked in order of magnitude of the third column – direct and indirect taxes as percentages of GDP in 1976.
[2] Income tax, social security contributions, and taxes on goods and services.
[3] Income tax and social security contributions.
[4] 1975 data.
Source: *Revenue Statistics of OECD Member Countries, 1965-1976*
Source: *Social Trends 10* (1980) table 6.15.

CHAPTER 3

THE ACCURACY OF INFORMATION

3.1 APPROXIMATIONS

Perfect accuracy in statistical information is possible only in limited circumstances. 'One plus one equals two' is perfectly accurate, but in applied areas of activity where problems have to be solved and decisions made, perfect accuracy of this kind is impossible.

Approximations are everywhere in statistics:

For example: export figures in the UK are based, not on actual sailings of ships with cargo, but on exporters' returns received by the Department of Trade during the year. Therefore the figures depend on the accuracy of the returns. At various times (such as in the early 1970s), it was discovered that British export figures had been underestimated by millions of pounds a month because small exporting companies had not made the necessary returns.

Another example: the instructions on some bottles of aspirin state: 'Take 1 or 2 tablets 3 or 4 times a day'. One tablet, 3 times a day would be 3 tablets in total; 2 tablets, 4 times a day would be a total of 8 tablets. The level of tolerance here is considerable. The instructions could be rewritten to say 'take between 3 and 8 tablets at intervals during the day depending on how you feel'. That is as accurate as they need to be. On the other hand some medicines have to be taken with so much accuracy that they can be applied only under the controlled conditions of a hospital ward.

Yet another example: a machine operator may be asked how long he spent maintaining his machine during recent weeks. His answer might be 'five minutes' one week, 'half-an-hour' another week and 'just over an hour' the third week. In fact the times might have been 'three minutes', 'thirty-two minutes' and 'sixty-eight minutes' respectively, but the approximations make more sense in answering a general inquiry which does not require a high degree of accuracy. The answer might have been different depending on the reason for the inquiry and who was making it.

Statistics is built upon approximations of one kind or another. There is very rarely an exact result to be derived from a statistical investigation. To the extent that statistics can be described as a social product rather than an objective method it is concerned with the level of accuracy required at a particular time. The accountant keeps track of every penny because both sides of the ledger, debit and credit, must balance to the last penny. The statistician may round figures to the nearest unit, or hundred or thousand depending on the level of accuracy required in a piece of information in order for it to be useful in solving problems and making decisions.

Even in areas where accuracy appears to be little problem it may be difficult to obtain. All scientific measurements are to some degree inaccurate, either because no measuring device can record an exact reading or because of the experimenter's inability to read the index accurately. The length of a line, for instance, can never be measured exactly even though it may normally be expressed as an exact number or fraction of a number of centimetres or inches. If the width of the marking line on the ruler is 0.002 centimetre, it follows that what is read off as one centimetre may be anything between 0.999 and 1.001 centimetres. For many purposes however, it will not make any practical difference if what is called one centimetre is in reality 1.001 centimetres.

A carpenter may aim for very high levels of accuracy in measurement when making joints for furniture, but a lower degree of accuracy when making similar joints for a roof.

Even exact mathematical results may be only approximate in relation to reality, since they are based on assumptions of a continuity of the circumstances assumed to exist for the purpose of the calculation. If the circumstances alter, so may the true value alter.

For example: by the time the national census figures are available, the demographic population statistics will have altered because of the births and deaths which have occurred in the intervening period. In fact, although at a specific moment the population of a country is a finite figure, the actual figure available will always be an approximation because of the impossibility of counting every single person alive at that moment.

3.2 DEGREES OF TOLERANCE

The degree of accuracy required in statistics depends upon the type of data being measured and the uses to which it will be put. For every measurement there will be a level of tolerance beyond which the inaccuracy becomes unacceptable, and within which the inaccuracy is acceptable.

For example: in estimating the size of an audience at a meeting, to say that there were 100 people, when there were in fact 90, might be acceptable, but to say that there were 50 or 150 might not be acceptable.

Another example: if someone is measuring a car they are thinking of buying to see if it will fit into their garage, an inch or two might not make any difference, but 5 or 6 inches might mean that the garage doors will not close. The level of tolerance is say 2 inches or 5 centimetres. The prospective buyer needs to know the length of the car to this level of accuracy.

Statistical tables usually contain approximations because tables are a summary of the collected data and are unlikely to contain all the information that was in the original survey. In summarising the data rounding is almost certain to take place. In most cases this is indicated by putting the degree of approximation in brackets:

For example: '(thousands)' would indicate that the column of figures is rounded to the nearest thousand. If the first figure in the column is 31,000, then the true figure would lie between 30,500 and 31,500.

Close approximations are often good enough for the purposes for which the data is required; statistics help to define the limits within which such approximations function. It is important that approximations should be clearly identified. In the same way inaccuracy must be allowed for in statistical investigation by recognising that it exists and that it can be kept within tolerable limits.

3.3 ERROR

'Error' means the difference between what is acceptable as a true figure and what is taken for an estimate or approximation.

Error in this sense does not mean a mistake. In the calculation $5 + 2 - 1 = 4$ the answer is wrong and therefore in mathematical terms it could be described as an error or a mistake. In statistical terms an 'error' is the difference between the approximate figure and the true figure.

For example: the size of a crowd at an open-air meeting may be estimated to be 20,000. Nobody knows exactly how many people attended the meeting either because nobody counted exactly or/and because nobody had been able to count the numbers because of constant movement with people arriving and leaving. However, at a particular moment there would have been an exact number in the crowd. If everybody had been made to stand still for long enough it would have been possible to count the numbers there. This would have been the 'true figure'. If this had been say 18,000, this figure could be compared with the approximate or estimated number of 20,000. The difference, 2000, is the error. This could be put in a different way by saying that the error was 10%.

The error in this example is not a 'mistake'. The size of crowds at open-air meetings are often estimated because of the cost, time and difficulty of arriving at an exact figure. Also, exact figures are not usually required.

However, it is important that the crowd size quoted in this example should be qualified in one way or another to indicate that it is an estimate and not an exact figure. The crowd estimate could be written as 'about 20,000', 'estimated to be 20,000' or '20,000 approximately'.

A mistake is usually involuntary and something to be avoided, while aporoximations are made to improve the presentation of figures and in this sense statistical error is deliberate. Error is a way of qualifying the degree of accuracy of a result (see also Section 11.5).

Most statistics reflect the degree of accuracy required rather than the degree of accuracy that is possible:

For example: if it had been thought necessary to measure the size of the crowd at the open-air meeting to an accuracy of say 2%, then a barrier could have been erected, turnstiles installed and everyone counted as they entered the meeting. This would be expensive and time consuming and there would need to be a good reason to make it worthwhile.

Even if there is an incentive to collect accurate data there may still be some degree of error involved. Crowds entering football grounds do have to pass through turnstiles because they have to pay and the numbers have to be counted because the stadium will have a limited capacity. So figures are issued for crowd sizes at football matches which could be completely accurate. In fact even under these conditions there is likely to be some error involved. The counting might be inaccurate at the turnstiles through human or mechanical failure, and some people may have climbed over the walls.

When the population of the UK is said to be 56 m., this does not mean exactly 56 m., but is more likely to mean 56 m. to the nearest million. This means that the population could be anywhere between 55.5 m. and 56.49 m. The figure of 56 m. could be as much as half a million people different from the 'true' figure. Therefore the 56 m. could be written as 56 m. $\pm \frac{1}{2}$ m. The $\frac{1}{2}$ m. is the degree of error involved in the approximation.

The widespread use of approximations means that any conclusions drawn from figures are themselves subject to error; therefore it is useful to know and to be able to identify the main types of approximations used.

3.4 ROUNDING

(a) To the nearest whole number

Fractions and decimals are frequently rounded 'to the nearest whole number'. The convention is to round 0.5 and above to the next highest whole number and 0.499 (recurring) to the next lowest whole number.

Therefore: 6.5 would become 7
6.499 would become 6

In the same way 65 would become 70 if rounded to the 'nearest whole ten' and 64.99 would become 60. Also, 650 would become 700 rounded to the 'nearest hundred' and 649.99 would become 600.

Survey figures are frequently rounded, because if some of the data is not very accurate there is little point in recording other figures with great accuracy, unless they are in some way independent.

For example: in a survey of petrol sold at service stations, recorded figures may be rounded to the nearest full litre or gallon or tens of litres or gallons, on the basis that there is likely to be some inaccuracy due to such factors as spillage and evaporation. This means making a decision as to how to round the actual figures. The choices are:

(i) To round up to the next highest whole unit, say to the next ten litres. An actual figure of 123 litres would become 130 litres recorded on the survey form.
(ii) To round down to the next lowest ten litres. An actual figure of 123 litres would be recorded as 120 litres.
(iii) To round to the nearest ten litres. An actual figure of 123 litres would become 120, and an actual figure of 125 litres would become 130 litres.

Rounding up (i) and rounding down (ii) are not recommended in statistics without very good reasons, because they give rise to biased error (see Section 3.6). If it was felt that there was always a wastage of petrol in the above example, so that the amount sold was always less than the amount recorded as going through the petrol pumps, there might be a justification for rounding down. However, rounding to the nearest unit usually means that the small amounts added tend to balance out the small amounts subtracted, so that the final total is close to the true total.

Very large figures become more comprehensible if rounded. To be told that the population of the UK is about 56 m. is easier to comprehend than to be given the figure 55,873,451, even if there was some way of knowing that the second figure was accurate. Even if this figure was accurate usually it would be unnecessary to quote it in full in most statistical presentations.

(b) **To the nearest even number**
Another procedure which can be used in an attempt to reduce bias from rounding is to round a number so that the digit preceding the final zeroes in the approximate value is even and not odd.

For example: if 125 is to be rounded to two significant figures (see (d) below), it would be rounded down to 120, while 135 would be rounded up to 140.

In most circumstances this procedure does not have any advantages over rounding to the nearest whole number.

(c) By truncation

This is similar procedure to rounding. It consists of the omission of the unwanted final digits.

For example: 15.268 truncated to four figures becomes 15.26; to two figures it would become 15.

In some currencies banks ignore very small denominations of coinage (like the halfpence) and pocket calculators truncate any digits lying outside their display capacity.

This procedure produces a downward bias into the results obtained.

(d) By significant figures

This is a rounding process by which the number of digits that are significant are stated and after that number zeroes replace other digits.

For example:

Calculated figure	Four significant figures	Three significant figures	Two significant figures
213.73	213.7	214	210
0.003726	0.003726	0.00373	0.0037
2,482,731	2,483,000	2,480,000	2,500,000
30,000	30,000	30,000	30,000
20,518	20,520	20,500	21,000

Notice that the zeroes which only indicate the place value of the significant figures (tens, hundreds, thousands) are not counted as significant digits.

Frequently, significant figures are the digits that carry real information and are free of spurious accuracy (see Section 3.10).

3.5 ABSOLUTE ERROR

This is the actual difference between an estimate or approximation and the true figure.

For example: a housewife may expect to spend £10 on her shopping, but actually spends £12.50. The absolute error is £2.50. Another housewife expects to spend £20 on her shopping, but actually spends £22.50. Again the absolute error is £2.50.

However, it is clear that in this example, the second estimate is better than the first in the sense that in the first case the error was 25% of the original estimate, while in the second case the error was $12\frac{1}{2}\%$. For a comparative measure, relative error is used.

3.6 RELATIVE ERROR

This is the absolute error divided by the estimate, often expressed as a percentage:

$$\text{Relative error} = \frac{\text{absolute error}}{\text{estimated figure}} \times 100$$

Therefore: $\frac{2.50}{10} \times 100 = 25\%$

$\frac{2.50}{20} \times 100 = 12\frac{1}{2}\%$

3.7 BIASED, CUMULATIVE OR SYSTEMATIC ERROR

If in a series of items the errors are all in one direction, the result will be a biased, cumulative or systematic error.

For example: if people are asked to give their ages at their last birthday, the total age of the group will be lower than the real total:

	Ages	
Actual		Approximation
Years	Months	(age last birthday)
15	10	15
18	2	18
17	7	17
18	5	18
70	0	68

The absolute error is 2 years

The relative error is $\frac{2}{68} \times 100 = 3.03\%$

The aggregate error is normally much larger in biased error than the aggregate unbiased error.

3.8 UNBIASED OR COMPENSATING ERROR

This is when the approximation is to the nearest whole number or complete unit.

For example: if people are asked to give their ages to their nearest birthday the total age of the group is likely to be similar to the real total.

	Ages	
Actual		Approximation
Years	Months	(age at nearest birthday)
15	10	16
18	2	18
17	7	18
18	5	18
70	0	70

In this case there is no absolute or relative error because the small amounts added and subtracted balance out so that the final (approximated) total is the same as the true total.

This will not always work out exactly as it does in this example, but generally the unbiased error will tend to be small and the greater the number of items the smaller the error will tend to be.

3.9 CALCULATIONS INVOLVING APPROXIMATION AND ERROR

In both theory and practice it is necessary to make calculations in which approximations and errors are present. This is because so much statistical data is rounded or is accurate to a certain number of significant figures or contains an element of biased or unbiased error. Much of the time the fact that approximation is present in data does not matter, because the degree of accuracy falls well within the levels of tolerance of the material being used and the needs of the subject being studied. However, it is useful to know what influence approximations can have on data and on the results of calculations.

For example: many population statistics are available only in an approximate or rounded form. Birth rates and death rates are accurate to the nearest whole number. Therefore in making calculations about a town's future population and commercial development and building based on it or the building of schools and the development of cemeteries, the approximate nature of the statistics has to be taken into account.

If it is known that the population of a town is 120,000 to the nearest thousand and the birth rate is 15 births per thousand, accurate to the nearest whole number; it is possible to calculate the limits of error in the statistics:

120,000 to the nearest thousand can be rewritten as 120,000 ± 500.
15 to the nearest whole number can be rewritten as 15 ± 0.5.

The greatest possible birth rate would then be:

120,500 × 15.5 = 1,867.75 births

The lowest possible birth rate would be:

119,500 × 14.5 = 1,732.75 births

All that can be said is that the number of births in the town is likely to be between these two figures:

1800.25 ± 67.5

The difference of 135 births between the two figures could be an important factor in future planning, although for some plans the difference could be considered unimportant.

Another example: in setting economic growth targets governments frequently assume various possible growth rates of a 3% growth, plus or minus 1% form, because of the difficulty of predicting growth more accurately than this.

The difference between a 2% rate of growth and a 4% rate of growth is considerable, particularly over a period of time.

Yet another example: manufacturers are frequently working on estimates (or approximations) of the costs of their raw materials, as well as estimates of their labour and other costs. They have to estimate maximum and minimum prices for their products, making assumptions about their costs and possible profit: a car component manufacturer may know that his fixed costs are £1 per component, but the cost of the materials in the component, which are at present £2, could rise or fall by 10% in the next year. His wage costs of £3 per component could rise or fall by 15% depending on levels of productivity and wage increases. Therefore during the next year his total cost per unit would be:

$$£1 + (2 \pm 10\%) + (3 \pm 15\%)$$
$$= £1 + (2 \pm 20p) + (3 \pm 45p)$$
$$= £6 \pm 65 \text{ pence}$$

Therefore he would not be sure of making a profit unless he sold his components for at least £6.65 a unit. However, if everything went well and raw material prices fell and productivity rose much faster than wage rates, the manufacturer could break even at a price of £5.35.

These examples indicate the need to be aware of the approximate nature of many statistics in making calculations. The basic methods of making calculations of levels of approximations are:

(a) **Addition**

(i) add 17 ± 0.5 and 3 ± 0.01

The highest possible result is:
$$17.5 + 3.01 = 20.51$$
The lowest possible result is:
$$16.5 + 2.99 = 19.49$$

Total $\quad$ 40.00

The mid-point is 20
The limits of error (the difference between 20.51 and 19.49) is ± 0.51
Therefore $(17 \pm 0.5) + (3 \pm 0.01) = 20 \pm 0.51$

It can be seen that the error in the aggregate is the sum of the absolute errors in the component parts $(0.5 + 0.01)$.

(ii) 20,000 correct to the nearest 1000 plus 4700 correct to the nearest 100

This can be rewritten:
$(20,000 \pm 500) + (4700 \pm 50)$
The maximum result is $20,500 + 4750 = 25,250$
The minimum result is $19,500 + 4650 = 24,150$
The mid-point is 20,000
The limits of error $= \pm 550$
Therefore $(20,000 \pm 500) + (4700 \pm 50) = 24,700 \pm 550$

(b) **Subtraction**

Subtract 20 ± 2 from 100 ± 10

The maximum result is the highest figure minus the lowest:

$110 - 18 = 92$

The minimum result is the lowest figure that can be obtained from 100 ± 10 minus the highest figure which can be obtained from 20 ± 2:

$90 - 22 = 68$

The mid-point $80 \left(92 + 68 = \dfrac{160}{2}\right)$

The limits of error $= \pm 12$

Therefore $(100 \pm 10) - (20 \pm 2) = 80 \pm 12$

It can be seen that the error in the answer (12) equals the sum of the errors in the individual parts (2 and 10).

When adding or subtracting with rounded numbers it is important to remember that the answer cannot be more accurate than the least accurate figure.

For example:

Add 327 to 631 and 700, where 700 has been rounded to the nearest 100

$327 + 631 + 700 = 1658$

But since the least accurate figure is to the nearest 100, the answer must be given to the nearest 100 and therefore the answer will be 1700.

Any attempt to be more exact can only result in spurious accuracy (see Section 3.10).

(c) **Multiplication**

Multiply 100 ± 2 by 20 ± 1

The maximum possible result is obtained by multiplying the highest figure that can be obtained from 100 ± 2 by the highest figure that can be obtained from 20 ± 1.

$120 \times 21 = 2142$

The minimum possible result is obtained by multiplying the lowest figure that can be obtained from 100 ± 2 by the lowest possible figure that can be obtained from 20 ± 1.

$98 \times 19 = 1862$

The mid-point is $2002 \left(\dfrac{2142 + 1862}{2}\right)$

The limits of error $= \pm 140$

Therefore $(100 \pm 2) \times (20 \pm 1) = 2002 \pm 140$

(d) **Division**

Divide $1000 \pm 2\%$ by $100 \pm 1\%$

This can be rewritten: 1000 ± 20 by 100 ± 1

The maximum result is obtained by dividing the minimum figure that can be obtained from 100 ± 1 into the maximum figure that can be obtained for 1000 ± 20.

$$1020 \div 99 = 10.303$$

The minimum result is obtained by dividing the maximum figure that can be obtained from 100 ± 1 into the minimum figure that can be obtained from 1000 ± 20.

$$980 \div 101 = 9.703$$

The mid-point = $10.003 \left(\dfrac{10.303 + 9.703}{2} \right)$

The limits of error = ± 0.3

Therefore $\underline{(1000 \pm 2\%) \div (100 \pm 1\%)}$
$\underline{ = 10.003 \pm 0.3}$

The important factor to remember in these calculations is that the greatest and least approximations are possible and to make allowance for this fact when these figures are used.

In fact in business, bias may be used purposely in certain circumstances. In estimating future expenditure it may be prudent to base the estimate on biased error to produce a higher figure than an unbiased result.

In estimating individual income it may be useful to use the maximum result of an estimate for some purposes and the minimum result for other purposes. It may be decided that earnings will be about £500 a month (the 'about' meaning say ± £40). This means that earnings could be between £460 and £540 a month or £5520 to £6480 per year, an annual difference of nearly a thousand pounds. For tax purposes the lower figure might be estimated, while in order to obtain a mortgage the higher figure might be used. Adjustments would have to be made when the true figure is known at the end of the year.

3.10 SPURIOUS ACCURACY

It is not sufficient to appreciate that complete accuracy is usually impossible, also it is important that claims are not made for such accuracy where it does not exist.

For example: to write 6.354 means that an accuracy of up to three decimal places is being claimed. It must mean that, for if the accuracy is only to two decimal places then the '4' is a guess and it is pointless to include it.

When a figure implies an accuracy greater than it really has, such accuracy can be termed spurious.

It is easy to be misleading with statistics:

For example: 'buy now and save 100%', may mean that there has been a reduction of 50% from say 10p per unit to 5p. A 100% reduction would mean giving the commodity away free, unless the 100% was based on something else not mentioned (like 100% of a future price increase).

Another example: to arrive at a high wage it is possible for an employer to add up wages. If one basic working hour at £2 is added to one overtime hour at £3 and one double-time hour at £4, it is possible to arrive at an average hourly wage of £3 (that is £9 divided by 3). In fact because most of the working hours are basic hours, the average is in fact likely to be near £2.

Dr Johnson is supposed to have said: 'Round numbers are always false.' It would be perhaps equally true to say that 'exact numbers are always false'.

For example: a commodity is reported to have been sold in the USA for £21,333. In fact the original report may have been that it was sold for about 48,000 dollars, this being an approximation of the exact sale price of $48,850. The figure of £21,333 was arrived at by dividing the $48,000 by the day's approximate exchange rate of $2.25 to £1. Therefore an approximate figure has been converted using an approximate conversion rate to produce an exact figure in sterling.

This is spurious accuracy. It would have been more accurate to say that the commodity had sold for about £21,000, or between £21,000 and £22,000.

The statement: 'lies, damn lies and statistics' arises from the use of spurious accuracy as well as the careful selection of figures to support a particular argument. Perhaps all statistics need to have a label attached to them: 'treat with care and understanding'.

ASSIGNMENTS

1 Discuss the importance of accuracy in statistics. What is meant by spurious accuracy?

2 Company A produces one commodity. The cost per unit of this commodity is made up of:

(i) labour costs of £3 (to the nearest £) per unit,
(ii) raw materials costing 90p (to the nearest whole 10p) per unit,
(iii) fuel and power costing an average of 10.3p (to three significant figures) per unit,
(iv) overhead costs averaging 85.2p (to one place of decimal) per unit.

What is the maximum and minimum cost per unit of output for this commodity?

3 Find examples of rounding in a variety of publications (newspapers, magazines, journals, company reports, government publications). Discuss

the degree of rounding in these examples and the results of this rounding on the interpretation of the data.

4 The population of a town is estimated to be 196,000 when the actual population is 200,000.

In these figures, what is the:

(i) absolute error,
(ii) relative error,
(iii) percentage error?

5 Discuss the following terms used in statistics:

(i) degrees of tolerance
(ii) error
(iii) rounding
(iv) absolute error
(v) relative error
(vi) biased error
(vii) unbiased error

CHAPTER 4

COLLECTING INFORMATION

4.1 SURVEYS

If data is not already available but it is needed to help solve a problem, it has to be collected. This is primary data. The collection of all primary data involves carrying out a survey or inquiry of one type or another.

A very limited survey or inquiry can be carried out in a few minutes by observation.

For example: how well used is the firm's car park? By looking out of the window it may be observed that there are fifty parking spaces and that five of these are empty. Therefore the car park has 90% usage.

This is a limited survey given spurious validity by using a percentage. The one rapid observation gives a result only for that particular time of day on that particular day. This could be described as a sample of all possible observations of the car park all day every day. However, it is such a small sample (see Chapter 5) that it is not likely to be very accurate or much help in deciding whether the firm needs a larger car park or could take over part of the car park for some other use.

Two observations would provide a check on each other and a series of systematic observations would provide a survey with a good chance of providing a useful answer to the question: useful in the sense of helping to make a decision. In the example, argument over whether the car park is fully used or under used can be narrowed by carrying out a statistical survey. The survey will provide statistical evidence which can help agreement to be reached.

The larger and more detailed a survey the more chance there is of it being both valid and accepted. However, it is better to carry out some type of survey, even if it is very limited, rather than none at all. One rapid observation of the car park is perhaps better than none at all.

For example: it has been suggested that one way of very quickly testing the morale of staff in offices, factories, schools and colleges is to walk into the building and count the number of people smiling. The more smiles, the higher the morale and job-satisfaction.

This 'smile test' should not be taken too seriously! What is required to produce useful results is a systematic survey.

Surveys produce primary data which is collected from basic sources in order to satisfy the purposes of a particular inquiry. Secondary data often provides the framework of information, leaving matters of relative detail to be filled by special surveys.

Survey statistics are available on a wide range of topics:

(a) **Government surveys** form an important part of the total.
(b) **Market research surveys** are in the main carried out for a particular client and are not published in any form.
(c) **Research surveys** are carried out by academics and other research workers often with the results published in journals.
(d) **Trade associations** collect statistics on sales for their members to help establish their share of the market.
(e) **Firms** commission *ad hoc* surveys on a wide variety of subjects.

Given the limitations of time and costs there are few limits to the nature or quality of data that can be obtained by means of a survey.

4.2 SURVEY METHODS

Whether a survey is a 100% survey of all possible items or is a small sample survey, there will be a series of stages in carrying it out:

(a) The survey design
This depends on the subject of the survey, what methods are available and the amount of time and money that can be spent collecting information. The UK Census of Population is unusual, because it is carried out on a 100% basis, aiming to find out a very wide range of information from every household in the country. On the other hand the *Family Expenditure Survey* is concerned mainly with households' income and expenditure and it is based on a sample. The *FES* employs a variety of methods so that the design of the survey is fairly complex (for a description see Section 5.19).

The survey design includes decisions on the size of the sample and the type of sample to use as well as decisions on the method of collecting the information.

(b) The pilot survey
This is a preliminary survey carried out on a very small scale to make sure that the design and methodology of the survey are likely to produce the information required. The survey may be tried out on two or three people

instead of two or three thousand. It is then possible to alter the design of the survey if it is discovered to be inadequate, before time and money have been spent on the main survey.

(c) **The collection of information**
The main methods of collecting primary data are through observation, interview and questionnaire. Most surveys use a combination of these methods (see Sections 4.3, 4.4, 4.5).

(d) **Coding**
It is useful in processing survey forms to pre-code the questions so that answers can easily be classified and tabulated. Coding can mean numbering or lettering questions or include more elaborate methods of identifying groups of answers.

(e) **Tabulation**
Once the information has been collected it has to be classified and tabulated (see Sections 7.2 and 7.3). In designing a survey it is useful to consider the problems that might arise at the tabulation stage.

(f) **Secondary statistics**
The information contained in the tables will often need to be summarised by calculating secondary statistics such as percentages and averages.

(g) **Reports**
The final stage of a survey is usually to write a report on the results and to illustrate the results with graphs and diagrams (see Section 7.5 and Chapter 8).

There are a number of methods of collecting primary data. They are often used together in a survey but it is easier to consider them separately:

4.3 OBSERVATION

Observation is the method used by consumers to compare prices and by companies to gather information on the use of their car park or to inspect the quality of their products.

Direct observation can be used to discover a variety of types of information including aspects of social and economic behaviour. It has been used to look at consumer behaviour, working methods and a range of social activities. In many cases the observer will try to be as unobtrusive as possible in order not to directly participate in the events being observed. This may be difficult unless the observer is hidden.

In time and motion studies the investigator has often been caricatured as a man with a stop-watch, notebook and binoculars, trying to watch people working without being seen. The implication is that if he was seen the workers might work harder than usual. In the same way an investigator sitting in a classroom to observe the behaviour of the children or the teacher could well influence their behaviour. The children might behave better than usual because they want to impress the stranger; or they might behave worse because they want to show off.

In an attempt to avoid these problems there have been approaches on two extremes:

(a) **Participant observation** in which the observer becomes a participant in the activity being observed. The observer works in the factory to observe how hard people work, or takes part in classroom activity to observe the behaviour of children and teachers. This approach is very time consuming, and the observer does not know the extent of his own influence on what is happening.

(b) **Systematic observation** (or objective observation) is used at the other extreme to observe only events which can be investigated without the participants knowing. An example is road use. Observers can watch a section of road and note the number and type of vehicles at various times of day on different days. This method is very objective, but the motives of the people observed are not questioned. It is not known why the drivers are travelling on that particular road at that particular time. To overcome this problem, drivers can be asked the purpose of their journey. This may still be systematic but will not be objective in the sense that the participants will now know what is happening.

It is open to question how far people should be observed without their knowledge and at what point this becomes an invasion of privacy. Observation has been applied to methods of changing people's behaviour and it is a matter of opinion how far this should go.

For example: it was hoped that background music in supermarkets would increase the level of activity and either encourage people to buy more or to shop faster. Observation showed that when loud music was played rather than soft music, sales did not change but people did shop faster.

(c) **Problems of observation**

(i) *Objectivity* – to remain objective the observer cannot ask the questions which will help him to understand the events he is observing.
(ii) *Selectivity* – the observer can be unintentionally selective in perception, recording or reporting.
(iii) *Interpretation* – observers may impute meanings to the behaviour of people which the people themselves do not intend.

(iv) *Chance* – a chance event may be mistaken for a recurrent one.

(v) *Participation* – observers can influence events because people realise that they are being observed and change their behaviour.

(d) **Mechanical observation** is used in survey methods under certain conditions. The number of vehicles passing a particular point on the road can be recorded mechanically. The main problem is that it is difficult for mechanical means to distinguish between types of vehicle.

Sophisticated mechanical means are used to provide more complex information, included means such as television, film and tape recorders. The use of these methods can be expensive and can provide the survey team with too much unselected information.

Mechanical means of observation enable much more detailed material to be collected than would be possible by an observer working alone. Since the information is recorded it can be analysed in depth some time later. Also, using a piece of apparatus may avoid the influence of the observer on events, although people may still be influenced because they know they are being checked. Meters are used on television audience measurement to record the length of time the set is switched on and the channel to which it is tuned. Mechanical methods of inspection and testing quality are commonplace in industry (see Section 11.8).

Direct observation is the classic method of scientific inquiry; biologists, physicists, astronomers and other natural scientists rely for their accumulated knowledge on centuries of systematic observation. In the social sciences it can be used as a method of watching humans as if observing animals, in a detached, relatively objective way. Other methods of collecting primary data tend to be less objective.

4.4 INTERVIEWING

An interview can be described as a conversation with a purpose.

In an informal sense everybody uses interviewing to obtain information. 'What was the score in the football match last night?' If the answer consists only of the score and the questioner wants to know more about the game, he may have to ask more questions. 'Who scored?', 'When were the goals scored?' 'How did player X play?', 'Was there a large crowd?' and so on.

In many cases questioning will start with general questions, or alternatively with questions that are easy to answer. More specific and more complex questions can follow up certain points to clarify them and provide a greater understanding of the subject.

Similar methods are used in more formal interviewing. A 'formal' interview is a conversation between two people that is initiated by the interviewer in order to obtain information. The interview is likely to be more structured than the usual conversation, because the interviewer will present each topic by means of specific questions and will decide when the conversation on a topic has satisfied the objectives of the interview.

Interviews are used in a wide variety of circumstances and for many purposes:

(a) **Attitudes**: opinion polls are based on interviews to discover people's attitudes towards a proposed product that a company is developing, or to test people's political views, or to measure the change made to people by particular events (such as an advertising campaign).

(b) **Motives**: interviews are used in an attempt to discover why people are behaving in a particular way. Road usage surveys include interviewing drivers to ask them where they are going and why they are making their journey.

(c) **Job selection**: more or less formal interviews are used for selecting employees.

(d) **Reporting**: interviewing has reached a sophisticated level on radio and television.

The great advantage of interviewing over direct observation is that it is possible to question a person's motives and attitudes, about changes in behaviour and about possible future behaviour.

Interviews may be:

(a) **Formal**: where set questions are asked. Surveys on road use tend to be formal, with all drivers being asked the same list of questions.

(b) **Informal**: where the questions may follow a pattern but will vary between interviews in order and content. Job interviews are often relatively informal with similar questions being put to each candidate, but varying in response to the answers received. Some employers use two interviews for their recruitment, one very informal, the other formal.

Interviews are not uniformly successful. Respondents differ in ability and motivation, interviewers differ in skill and experience, and the interview content differs in feasibility. Interviews are an interactive and subjective method of finding information. Formal interviewing attempts to reduce these influences. This has been carried to the extreme of erecting a screen between the interviewer and the respondent. Usually the two are face-to-face and they may take an instant liking or disliking to each other.

The way questions are asked may influence the answers. For example, it is not unusual for electoral canvassing returns for the same area of a constituency to show that both major parties have a majority of the voters pledged to vote for them. This 'impossible' situation can arise because a number of voters may want to give the answer most likely to please the canvasser, or may feel that by giving an expected answer the interview will be over quickly. This type of interview is often carried out in the evening, on the doorstep of the voter's house, with a favourite television programme on in the background. All of these factors influence the success of the interview measured in terms of the accuracy of the information received.

The success of an interview will depend very much on the two protagonists involved, the interviewer and the respondent.

(a) **The respondent**: the ability of the respondent to 'make a success' of the interview will depend on:

   (i) *The accessibility of information* – the information being sought by the interviewer has to be accessible to the respondent so that he has it clearly thought out and is able to express it in the terms used by the interviewer.
  (ii) *Role* – the respondent needs to be clear about his role particularly in the sense of knowing what information is relevant and how completely he should answer.
 (iii) *Motivation* – the respondent needs to be motivated to answer questions and to answer accurately. The interviewer can suggest why it is in the respondent's interests to answer the questions. Up to a point people are happy to give interviews because they are asked; for long interviews greater incentives may be required. People will often answer questions at length if they believe that they can influence events (such as the siting of a new airport, or the route of a new motorway). Payment may help to provide an incentive, but can encourage people to become respondents even when they know little about the subject.
  (iv) *Prestige* – if the interview is carried out for a well-known or prestigious company or institution or has government backing, respondents may be encouraged to answer questions carefully.

(b) *The interviewer*: the interviewer may influence the results of the interview in obvious ways, such as careless recording of answers, or poor reporting of results, or by cheating. It has been known for interviewers to fill in questionnaires themselves to save themselves the trouble of collecting the answers. There are more subtle ways in which the interviewer may influence the result, these are referred to as *interviewer bias*:

   (i) By the way questions are asked.
  (ii) By the extent of probing or asking supplementary questions and the kind of answers the interviewer expects.
 (iii) By the interviewer expecting that early replies of a respondent establish a pattern which will be followed in later answers.
  (iv) By the interviewer expecting a person of a certain age and appearance to answer in a particular way.

Added to the problems arising from respondent expectations and interviewer bias are problems common to any questionnaire (see Section 4.5). The organisation and sequence of questions may influence the results of the interview.

All these problems can be reduced by good selection, training and supervision of interviewers so that they are aware of the problems that can arise and try to avoid them. Market research organisations use experienced interviewers for their door-to-door and street-corner interviews.

Used carefully the interview is an excellent method of collecting quite complicated information. In practice the amount of information collected by interviewing is limited by time and cost.

4.5 QUESTIONNAIRES

A questionnaire is a list of questions aimed at discovering particular information.

Questionnaires can be distributed by hand (pushed through people's letter-boxes) or by post. Frequently they are used in interviews to provide the interviewer with a set list of questions to ask.

(a) The advantages of postal questionnaires

(i) they are relatively cheap to distribute;
(ii) therefore they can be sent to large numbers of people;
(iii) the answers can be carefully considered.

(b) The disadvantages of postal questionnaires

(i) it is not possible to explain questions or to follow them up;
(ii) a poor response is usual, unless there is a very strong incentive to return the questionnaire.

Questionnaires distributed by hand usually have a more limited distribution because of the costs involved. However, the response rate may be much higher because they can also be collected by hand, and this provides an incentive to complete the form.

In all questionnaires problems can arise from the design of the questions. There are a number of points that need to be considered in question design.

(c) Factors involved in question design

(i) questions should be simply and clearly worded so that they can be understood by the 'average' respondent;

(ii) questions must be clearly useful and relevant, and designed to produce the desired information;
(iii) questions must be free from bias;
(iv) questions must not be so personal and private that the respondent will be reluctant to give an honest answer;
(v) the order of questions needs to be logical and help the respondent to remember the answers;
(vi) questions should be unambiguous and should not include vague words, such as 'fairly' and 'generally';
(vii) leading questions need to be avoided; for example: 'don't you think something should be done about . . . ?', encourages a positive answer;
(viii) hypothetical questions are of limited value.

(d) **Factors involved in the design of the questionnaire form**

(i) it needs to be clear from the form who should complete it;
(ii) it needs to be clear where the answers should be recorded;
(iii) sufficient space should be available to complete the answers;
(iv) the questions need to be set out so that it is clear how they follow on;
(v) as many questions as possible should be able to be answered by 'yes' or 'no', or by the respondent deleting a word or phrase, or by ticking a box;
(vi) the convenience of both the respondent and the survey team who will process the questionnaires need to be considered.

Everybody fills in forms of one kind or another, for the Inland Revenue, for the VAT man, to enrol on a course, to claim social benefit, to register a car and so on. Many forms are poorly designed, by providing limited space for 'home address' or by including ambiguous questions. The fact that many forms are poorly designed or include poor questions indicates the difficulty of producing a well-designed form and clearly understood questions. The objective of producing a questionnaire is to collect information, therefore it is worth organising and designing it carefully.

4.6 COLLECTING INFORMATION

Primary data is collected by one or a combination of these methods. Questionnaires are often used with interviews, particularly formal interviews, and with systematic observation. Whichever method or combination of methods is used the important factor is to obtain accurate information.

ASSIGNMENTS

1 Select any form or questionnaire and analyse the good and bad points of its design.

2 Design a questionnaire for a company which is considering the development of its social and sporting facilities. The questionnaire is to be distributed to all the employees of the company in order to discover their attitudes to these developments.

3 Carry out a survey on the road usage of a stretch of road, or the use of a car park. Analyse the results of the survey and write a short report on these results and on the problems of carrying out the survey.

4 Interview five people to discover their attitudes towards a four-day working week for:

(i) other employees,
(ii) themselves.

5 Discuss the problems involved in collecting primary information.

6 Comment on the following:

(i) Pilot surveys
(ii) Coding
(iii) Secondary statistics
(iv) Interviewer bias
(v) Postal questionnaires

CHAPTER 5
HOW TO SAMPLE

5.1 WHY SAMPLE?

A sample is anything less than a full survey of a population; it is usually thought of as a small part of the population, taken to give an idea of the quality of the whole. The 'population' is the group of people or items about which information is being collected.

It may seem desirable to base decisions on complete counts or measurements of people and commodities. Anything less than this may be felt to include only part of the information and to be open to a high degree of error and approximation. In practice it is only in limited circumstances that a 100% survey can be completed. The UK Census of Population every ten years is exceptional because it is a 100% survey. Within a company it may be possible to carry out a comprehensive survey of the labour force on a variety of subjects.

However, for much general information about people's opinions and attitudes and about the quality of commodities it is often impossible to carry out complete surveys. A marketing manager might like to carry out a census of all housewives on their attitudes to his product; a works manager might like to inspect in detail every item coming off the production line. In neither case is this feasible; it would be impossible for every marketing decision or every aspect of controlling quality to be based on a complete census of millions of people or thousands of items. A complete survey is not only impracticable but also unnecessary. In many cases a sample is preferable.

5.2 ADVANTAGES OF SAMPLING

(a) **Cost**: it is cheaper to collect information from 2000 people than from two million. However, the cost per unit (or item, or person) may be higher with a sample than with a complete survey. More skilled personnel may be used, new costs may be added such as those involved in sample selection and in calculations of the precision of the sample results. Also, overhead costs are spread over a smaller number of units. In spite of these extra costs, samples are usually so much smaller than a complete coverage (often 10% or less) that total costs are likely to be very much less.

(b) **Time**: information is often required within a specified time, so that a decision can be made and action taken. A sample requires less fieldwork, tabulation and data processing than a full survey. Also, following up non-response and other problems is quicker with a sample than a full survey because there are fewer items.

(c) **Reliability**: a high level of reliability can be achieved because fewer units are surveyed in a sample than in a full survey and therefore resources can be concentrated on obtaining reliable information. Well-trained field staff can be employed, more checks and tests made and more care taken with editing and analysis. Respondents may be more willing to provide detailed information if they know that they form a small sample of the population. They may feel that because they are representing the population they should provide reliable information.

In fact absolute accuracy may not be required. The size of a sample can be adjusted so that the resulting accuracy is sufficient for a decision to be made (see Section 11.5). If a larger sample is taken, or a full survey, resources are being wasted.

(d) **Resource allocation**: by using samples it is possible to carry out several studies concurrently, and therefore use resources efficiently.

(e) **The test may be destructive**: some sample tests destroy the product. No products would be left unless samples were used (examples include light bulbs and TV tubes).

This last point illustrates the fact that the interest is not in the sample items except in so far as they may be used to draw inferences about the population from which they are selected.

For example: a market research team will want to draw inferences about 20 m. housewives, not about the 3000 actually interviewed; the quality control inspector is interested in the thousands of components produced each day, not the few that he has tested to destruction.

5.3 OBJECTIVES OF SAMPLING

(a) **Descriptive statistics**: the most common objective of sample surveys is to estimate certain population statistics or parameters (such as averages and proportions).

A sample is selected, the relevant statistic is calculated and this is used as an estimate of the population statistic. The statistic should be accompanied by a statement about the accuracy of the result in terms of standard error. Standard error is a method of indicating the variability of a sample statistic; for instance to show the extent to which a sample average deviates from the population average (see Section 11.5).

For example: the average hours per week of overtime worked by the employees of a large company is estimated by carrying out a 10% sample survey. The average overtime is calculated (say 4.5 hours per week), with a statement about the standard error involved (say 0.5 hours). This indicates that the sample average may deviate by half-an-hour from the population average.

(b) **Statistics of inference**: another use of sampling is to test a statistical theory about a population. A theory or hypothesis is held about a population, a sample survey is carried out and the results are then interpreted to test whether the results support or refute the theory (see Section 11.6).

For example: it may be thought that the average hours per week of overtime worked by the employees of a large company is five hours. A sample survey is taken to test this theory.

The underlying objective of sampling is to describe the population from which the sample is taken. If sample results are to be used for decision making it is very important to assess the reliability of these results.

For example: it would be unfortunate for a company to reorganise on the basis of a 10% sample if the particular employees questioned in fact represent a distinct minority of all the employees in the company.

5.4 THE BASIS OF SAMPLING

The possibility of reaching valid conclusions concerning a population from a sample is based on two general laws:

(a) **The law of statistical regularity**: this states that a reasonably large sample selected at random from a large population will be, on average, representative of the characteristics of the population.

For this law to work, the selection of the sample must be made at random, so that every item in the population has an equal chance of being included in the sample. Also, the number of items in the sample must be large enough to represent the whole population and to avoid undue influence on the average by extreme items. The larger the number of items selected, the more reliable the information will tend to be. For instance, if three people are asked which political party they support there could be equal representation for three parties or a 'minority' party might be supported by all three people, even if it is known that most people support

two parties. A sample of three hundred people would be likely to provide a better impression of the support for each party and a sample of three thousand people might represent the views of the population quite well.
(b) **The law of the intertia of large numbers**: this states that large groups of data show a higher degree of stability than small ones. There is a tendency for variations in the data to be cancelled out by each other. Taking a large number of items, it is unlikely that the variations in them will all move in the same direction. If, for example, the average length of nails coming off a production line is 2.5 centimetres, some of the nails will be a fraction longer than this and some a fraction shorter, so that on balance they will be the average length.

These laws are part of a mathematical theorem, the central limit theorem, and are ways of describing the theory of probability. The theory of probability is the basis of statistical induction, which is the process of drawing general conclusions from a study of representative cases (for a further discussion of this theory and the basis of sampling see Chapter 11).

5.5 SAMPLING ERRORS

Errors in a sample survey may arise from both sampling errors and from non-sampling errors or bias.

(a) **Non-sampling errors**: these are due to problems involved with the sample design. Many of them would arise with a full survey, but some of them are due specifically to sample design, including such factors as the choice of a sampling frame and sampling units (Section 5.7 below).
(b) **Sampling error**: this is the difference between the estimate of a value obtained from a sample and the actual value. A sample may show that the average weekly wage of a group of employees is £100, when the actual average is £110 a week. The sampling error is £10.

Sampling errors arise because even when a sample is chosen in the correct way (by random methods), it cannot be exactly representative of the population from which it is chosen. The degree of sampling error will depend on the size of the sample; the larger the sample the smaller the error (see Sections 5.7 and 11.5). This is not dependent on the size of the population. A population of 3 m. does not require a larger sample than a population of 300,000 or 30,000.

The important point about sampling error is that providing the sampling method used is based on random selection it is possible to measure the probability of errors of any given size. The total error in a sample arises from both non-sampling and sampling errors and cannot be substantially reduced unless both types of error are simultaneously controlled. There is no point in taking a larger sample in order to reduce the sampling error if there are design faults in the sample.

5.6 SAMPLE SIZE

The size of a sample (the number of people or units sampled) is independent of the population size. It does depend on the resources available and the degree of accuracy required. Other things being equal a large sample will be more reliable than a small sample taken from the same population.

Therefore the number of items sampled is a matter of judgement based on the variability in the population. A population which is known to be very variable (including numbers of people with different opinions or including units of many types) will require a larger sample to represent it than a population known to be very homogeneous.

For example: samples of political opinions in the UK have increased in size in recent years because it is felt that the electorate has become more variable and volatile in its opinions.

5.7 SAMPLE DESIGN

(a) Principles of sample design
(i) To avoid bias in the selection procedure.
(ii) To achieve the maximum precision for a given outlay of money and time.

(b) How to design a sample
(i) Decide on the objectives of the survey.
(ii) Assess the resources available.
(iii) Define the sample population and the sample unit.
(iv) Select a sample frame.
(v) Decide on a survey method.
(vi) Choose a sampling method.

(c) Elements in the design
(i) *The sampling population* is the group of people, items or units under investigation.
(ii) *The sample units* are the people or items which are to be sampled. These units need to be defined clearly in terms of particular characteristics. For instance, in a sample of 'vehicles', this term needs to be defined in terms of say motor cars, buses and commercial vehicles. This would exclude motor cycles, bicycles, tractors and other vehicles that use the road. In practice it may be found that motor cars, buses and commercial vehicles need to be defined very clearly in their turn.
(iii) *The sample frame* is the list of people, items or units from which the sample is taken. The sample unit is defined and then a suitable sampling frame is sought. It should be comprehensive, complete and up-to-

date to keep bias to a minimum. General examples of sampling frames include electoral registers, telephone directories, wage lists.

(iv) *The survey method* includes designing questionnaires and deciding how to distribute them or how to carry out interviews or observations.

(v) *Sampling methods* fall into two categories:
    random samples (simple, systematic, stratified),
    non-random samples (multi-stage, quota, cluster).

The type of sampling method chosen depends on the nature and purpose of the inquiry. It is difficult to assess the sampling error involved in non-random samples and therefore in using these methods there is usually an attempt to include some element of randomness.

5.8 BIAS IN SAMPLING

Bias consists of non-sampling errors, which are not eliminated or reduced by an increase in sample size. Bias may arise from:

(a) **The sampling frame**: if this does not cover the population adequately or accurately.

(b) **Non-response**: if some sections of the population are impossible to find or refuse to co-operate. There are certain groups that tend to be under-represented in many surveys, these groups include the very rich and very poor, young adults, working housewives.

(c) **The sample**: if the most 'convenient' sample is selected it may be biased and non-random. Examples are housewives in a particular shopping centre at a particular time of day, or the top box of components in a delivery.

(d) **Question wording**: poorly worded and ambiguous questions and interviewer bias may cause problems in sample surveys in much the same way as they do in comprehensive surveys.

(e) **The sample unit**: a personal element may enter into selection. Substituting one unit or person for another may introduce bias.

Any of these factors may cause systematic, non-compensating errors in a sample survey.

For example: in 1936 in the USA the *Literary Digest* carried out a huge sample of ten million individuals, yet its forecast of the result of the US presidential election was wrong, because:

(i) The sample was picked from telephone directories which did not adequately cover the poorer section of the electorate;

(ii) Only 20% of the mail ballots were returned and these probably came predominantly from more educated sections of the population.

The sample returns indicated that Franklin D. Roosevelt would be defeated, whereas in fact he was elected with one of the largest majorities ever recorded in the USA.

There are a range of sampling methods to choose from in carrying out a sample survey:

5.9 SIMPLE RANDOM SAMPLING

In random sampling each unit of the population has the same chance as any other unit of being included in the sample.

(a) Random numbers

To select each unit on a random basis a lottery method can be used. For large groups it is not possible to number or name every unit of the population and then pick them out of a hat. Therefore random numbers are used. These can be computerised (such as the system for selecting British Premium Bonds through the Electronic Random Number Indicating Equipment: ERNIE), or contained in random number tables.

These tables are constructed so that each digit from 0-9 is found an equal number of times in large sections of the table with no more repetitiveness than should properly occur by chance and with no tendency for the numbers to form repeating patterns.

For example: in random sampling for consumer market research, the Electoral Register is often used as a sampling frame. The electors for a polling district are listed according to name and address, and each elector is given a number. If a particular register has 5000 electors, each elector will have a number from 1 to 5000. Random numbers can be read off the table in groups: for instance, 2412, 8627, 0143. The first sample number would be elector 2412 on the electoral register, the second elector picked would be number 143. Random number 8627 would be ignored because the sample population is only 5000. This process can be continued until the number of people required for the sample is reached (say 500 for a 10% sample).

(b) Sampling with and without replacement

Unrestricted random sampling is carried out 'with replacement'. This means that the unit selected at each draw is replaced into the population before the next draw is made, so that a unit can appear more than once in the sample.

In sampling without replacement only those units not previously selected are eligible for the next draw.

In applied statistics it is assumed that sampling is without replacement (in a lottery a winning ticket is not usually replaced in the hat or box to allow it to win a second prize). Simple random sampling is sampling without replacement.

(c) **The main features of simple random sampling**
  (i) It is 'simple' as compared with more complex methods which are also random (such as systematic and stratified sampling). In the USA it tends to be referred to as 'simple' probability sampling.
  (ii) The technique of randomisation ensures the validity of the techniques of inference such as deciding the confidence with which results can be accepted (see Sections 11.6 and 11.7).
  (iii) Simple random sampling is the standard against which other methods are evaluated.
  (iv) It is suitable where the population is relatively small and where the sampling frame is complete.

5.10 SYSTEMATIC SAMPLING

This is a form of random sampling, involving a system. The system is one of regularity. The sampling frame is taken and a name or unit is chosen at random. Then from this chosen name or unit every nth item is selected throughout the list.

For example: if the sampling frame contains 100,000 names and a 2% sample is required, the 2000 names can be selected at regular intervals. The first is selected at random from the first 50 names (50 because 50 × 2000 = 100,000). If the 35th name is picked the names are selected at regular intervals to make up 2000 in all (the 35th, 85th, 135th, 185th . . .).

(a) **The advantages of systematic sampling**
  (i) It is sufficiently random to obtain an estimate of the sampling error (it is sometimes referred to as quasi-random sampling).
  (ii) The systematic approach facilitates the selection of sampling units.

(b) **The disadvantages of systematic sampling**
  (i) It is not fully random, because after the first point every remaining unit is selected by the fixed interval.
  (ii) There can be problems if particular characteristics arise in the list of names or units at regular intervals, which would create bias. For example, every 10th house on a list of addresses or houses might be a corner house with different characteristics to the other houses.

5.11 RANDOM ROUTE SAMPLING

This is a form of systematic sampling used in market research surveys. It is used mainly for sampling households, shops, garages and other premises in urban areas.

An address is selected at random from a sampling frame (usually the electoral register), as a starting-point. The interviewer is then given instructions to identify further addresses by taking alternate left- and right-hand turns at road junctions and calling at every nth address (shop, garage etc.) *en route*.

(a) **Advantages of random route sampling**
(i) There may be a saving in time because the interviewer is given clear instructions about choosing the respondent.
(ii) Bias may be reduced because the interviewer has to call at clearly defined addresses and is not able to choose.

(b) **Problems with random route sampling**
(i) It is only quasi-random because the element of selection can be strong and the characteristics of particular areas (poor and rich areas, for example) may mean that the sample is not representative.
(ii) The method is open to abuse by the interviewer, because it may be difficult to check that the instructions have been fully carried out without a costly and time-consuming repetition of the route. The interviewer may not record non-response and avoid certain premises (with fierce dogs etc.!).

5.12 STRATIFIED RANDOM SAMPLING

This is a form of random sampling in which all the people or items in the sampling frame are divided into groups or categories which are mutually exclusive (that is, a person or unit can be in one group only). These groups are called 'strata'.

Within each of these stratum a simple random sample or a systematic sample is selected. The results of the sample for each stratum are processed. If the same proportion (say 5%) of each stratum is taken, then each stratum will be represented in the correct proportion in the overall result. This eliminates differences between strata from the sampling error.

For example: in a marketing survey the sales of cigarettes in a variety of outlets may be investigated by dividing the retail outlets into strata. In a particular town or urban area shops may be divided into large, medium and small outlets and a simple random sample taken based on shops from each category. A clear definition of the strata is important so that there is no overlap between shops.

(a) **Advantages of stratified random sampling**
(i) It may provide a more accurate impression of the population where there are clear strata than other sampling methods.
(ii) This sampling design may be an improvement for certain populations on a simple random sample.

(b) **Disadvantages of stratified random sampling**
(i) If the strata cannot be clearly defined, the strata may overlap, reducing the accuracy of the results.
(ii) Within the strata, the problems are the same as for any simple random sample or systematic sample.

The characteristic of random, systematic and stratified sampling is that every individual has a known probability of being included in the sample. Therefore these methods can be called random or probability sampling.

Non-random or judgement sampling methods are used when random methods are not feasible. This may be:

(i) When searching for the people selected by random methods would be a long and uneconomic task.
(ii) When all the items in the population are not known and either there is no suitable sampling frame or it is incomplete.
(iii) When a random sample would involve expensive travelling for interviews.

These three categories coincide with the three most common non-random methods of sampling: quota, cluster and multi-stage.

5.13 QUOTA SAMPLING

In quota sampling the interviewer is instructed to interview a certain number of people with specific characteristics.

The quotas are chosen so that the overall sample will reflect accurately the known population characteristics in a number of respects. Quota sampling can be described as non-random but representative stratified sampling.

For example: interviewers may be told to interview, over a period of several days, fifty people divided into age and socio-economic groups, to ask them their opinions on a television advertisement. These groups may be divided in proportion to the numbers in the population. Therefore the instructions to the inverviewers could be to interview on the basis of Table 5.1.

Table 5.1 interview selection table for quota sampling

Age groups	Socio-economic groups	Numbers	
16-25	A/B	1	
	C	3	
	D/E	1	5
25-45	A/B	3	
	C	9	
	D/C	3	15
45-65	A/B	4	
	C	10	
	D/E	6	20
65 and over	A/B	2	
	C	6	
	D/E	2	10
	Total		50

This table highlights the fact that the more characteristics that are introduced the more difficult the interviewers task becomes. Already finding the individuals of the right age and socio-economic groups in the numbers indicated on the table may be difficult. If a further division was made, for instance into male and female, each of these numbers would have to be further subdivided.

In this example, the instructions to the interviewer may be to go to a particular shopping area at a certain time to interview people in the numbers indicated on the table. The first people encountered who fit the characteristics listed are interviewed.

(a) **Advantages of quota sampling**
 (i) It may be the only feasible method if the fieldwork has to be completed quickly.
 (ii) There may not be a suitable sampling frame to use a random sample.
 (iii) Administration may be easy because there are no non-responses (although people may refuse to answer).
 (iv) The costs may be less than other forms of sampling because the survey can be carried out rapidly. However, the greater the number

of characteristics and the wider the geographical spread the more expensive the survey will be.

(b) **Disadvantages of quota sampling**
  (i) It is not a random sampling method and therefore it is not possible to estimate the sampling errors.
  (ii) Interviewer bias may be important, because the interviewer has to choose the respondents. Some groups may be over-represented - the better educated, more articulate, people who happen to be out and about. Within the quota groups interviewers may fail to secure a representative sample of respondents - they may choose mainly 16-18 year olds in the 16-25 age group.
  (iii) Non-response may not be recorded because these are people who refused to be interviewed but are not on any sample list.
  (iv) Control of the fieldwork (interviewing) is difficult and the interviewers may have great difficulty in 'recognising' age and class groups.

It can be argued that the greatest defects in sampling are at the interview stage and in processing the data and therefore that the sample itself is a small source of error; therefore it can be argued that the disadvantages of quota sampling are outweighed by the advantages. The fact remains that unless a random element is introduced in the selection of the sample it is not possible to estimate the sampling errors.

5.14 CLUSTER SAMPLING

In cluster sampling (or area sampling) clusters are formed by breaking down the area to be surveyed into smaller areas, a few of these areas are then selected by random methods and units (such as individuals or households) are interviewed in these selected areas. The units are selected by random methods.

For example: a map of an urban area is divided by a grid and a selection of these areas is taken at random (see Figure 5.1).

Perhaps areas *D, G* and *J* are chosen. Every household in these areas can be interviewed, or systematic or random route sampling used. A team of interviewers can be sent to the selected areas so that the survey can be carried out quickly.

Cluster sampling is popular where the population is widely dispersed and it is easier to sample a cluster of people than a range of people or households over a wide area.

It is often used to survey the distribution and possible markets for consumer durables such as television sets and washing machines. Also it is used for quality control where batches of items are removed from the production line for testing and inspection.

Fig 5.1 *urban area grid for cluster sampling*

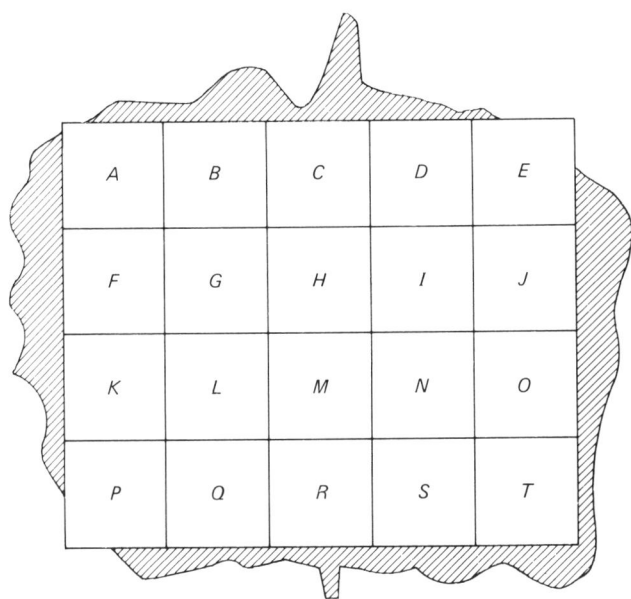

(a) **Advantages of cluster sampling**
(i) Where there is no suitable sampling frame this may be the only possible method.
(ii) Time and money is saved in travelling between locations and searching out respondents, because interviews are concentrated in a few small areas.

(b) **Disadvantages of cluster sampling**
(i) Clusters may comprise people with similar characteristics (areas *D*, *G* and *J* may all be relatively wealthy areas) and therefore the results may be biased (this can be reduced by taking a large number of small samples).
(ii) Although there are elements of random sampling in this method it is often difficult to estimate sampling errors.

5.15 MULTI-STAGE SAMPLING

This is a series of samples taken at successive stages:

(i) The country may be divided into geographical regions.
(ii) A limited number of towns and rural areas are selected in each region.
(iii) A sample is taken of people or households in the selected towns and rural areas.

Often the selection of towns and rural areas is carried out in such a way that the probability of selection is proportional to the size of the population. Therefore a town with a population of 120,000 would stand ten times the chance of being selected as a town with a population of 12,000. Correspondingly, an individual in the larger town stands one-tenth the chance of being selected as compared with an individual in the smaller town, so that individuals in both towns stand an equal chance of selection at the beginning of the sampling operations.

This may be the only practical method of sampling when the population of the whole country is being surveyed. Other methods would be too time-consuming and expensive. Regional differences can be allowed for by selecting regions at the first stage. At the second stage a random sample may be used, although again the proportion of urban against rural areas may be allowed for by a proportional distribution. At the third stage simple random or systematic sampling may be used to arrive at the individuals or households to be surveyed.

For example: the *Family Expenditure Survey* – this survey presents an excellent example of the multi-stage sampling method. It is a continuous survey introduced in the UK by the government in 1957 and conducted by the Office of Population, Censuses and Surveys. Each year a sample of over 10,000 households is selected:

(i) The country is divided into 1800 areas, from which a stratified sample of 168 areas is chosen. Stratification is by region, type of area (urban/rural) and by rateable value. This is to ensure that the areas selected are representative of the whole country.
(ii) Selected areas are divided into districts and four districts from each are selected by systematic sampling.
(iii) From each chosen district 16 households are selected by taking a systematic sample from the electoral register. The total number of households selected is therefore 16 (households) × 4 (districts) × 168 (areas) = 10,752.

Of these 10,752 households about three-quarters co-operate. These are visited and each member of the household over 16 is asked to keep records covering a period of two weeks of all payments made. In addition, the interviews complete questionnaires on each household covering items of regular expenditure such as gas and electricity bills, details of occupation, income, age and marital status. A payment of £1 is made to each member of the household involved with the survey.

All the documents are checked by the survey officials and then forwarded to the Department of Employment for analysis and publication. The results of this survey form the basis of the Index of Retail Prices (see Section 12.5).

There are a range of other sampling methods used under particular circumstance. These include:

5.16 MULTI-PHASE SAMPLING

This is a type of sample design in which some information is collected from the whole sample and additional information is collected from sub-samples of the full sample, either at the same time or later.

In a survey of households basic data may be needed from all the households. This may include such areas as the occupation of the head of household, household income, ages and so on. However some data may be required from only a small sub-sample because:

(i) The data is less important.
(ii) Some factors may be known to be constant in the population and therefore reasonable accuracy can be achieved from a smaller sample.
(iii) Some information may be so costly or troublesome to collect that it is only possible to survey a small sample.

This method of sampling can reduce costs and reduce the burden of work on respondents. It was used in the UK Census of Population in 1971 to collect more detailed information on people's qualifications in science and technology.

5.17 REPLICATED OR INTERPENETRATING SAMPLING

This is when a number of sub-samples, rather than one full sample, are selected from a population. All the sub-samples have exactly the same design and each is a self-contained sample of the population.

For example: a full sample of 500 employees could be divided into two sub-samples of 250 employees, or five sub-samples of 100 employees and so on.

(a) **Advantages of replicated sampling**
(i) If the size of the total sample is too large to permit the survey results to be ready when they are wanted, one or more of the replications can be used to obtain advanced results.
(ii) These samples can throw light on variable, non-sampling errors, because each of the sub-samples produces an independent estimate of the population characteristics. Therefore, if each sample is carried out by a different interviewer, it is possible to obtain an estimate of between-interviewer variation.

(b) **Disadvantages of replicated sampling**
The main problems can arise from the cost in time and money of carrying out a series of samples.

5.18 MASTER SAMPLES

These are samples covering the whole of a country to form the basis (that is, to provide a sampling frame) for smaller, local samples.

The US government has carried out master samples of agriculture to provide a framework for local agricultural surveys.

The units in the master sample must be fairly permanent or long lasting if the results are to be useful for any length of time.

5.19 PANELS

A groups of people is selected from a survey population by a random sample. They form a panel of people who are surveyed at various times over a period of time. This means that the same information is asked for from the same sample at different times.

For example: the attitudes of a panel of car owners can be surveyed before and after an advertising campaign. The results can be compared to see the influence of the advertising campaign.

(a) **Advantages of panels**
  (i) Changes and trends in behaviour and attitudes can be surveyed.
  (ii) The effects of specifically introduced measures can be estimated.
  (iii) The survey team can study those who change their views and those who do not.
  (iv) Evidence may be produced on the ordering of variables; which was cause and which was effect - an ordinary survey can show an association between a worker's attitude to his job and his position in the firm, but it does not indicate which came first.
  (v) Overhead costs can be spread over a period of time.

(b) **Disadvantages of panels**
  (i) It may be difficult to recruit the initial sample: people may not want to be involved with a panel.
  (ii) Panel mortality tends to be high - people leave panels for a variety of personal and other reasons, there is often a 50% drop-out over a few weeks.
  (iii) Panel conditioning means that the members may become untypical of the population they represent - TV panels start to view different

programmes or concentrate more on programmes because they know that they will have to answer questions on the programmes.

The sampling discussed in this chapter should not be confused with conscious or purposive sampling. This consists of taking carefully chosen samples to present the 'best' units, not typical or representative units.

ASSIGNMENTS

1 Analyse the advantages and disadvantages of at least five types of sampling methods.

2 Discuss the importance of the concept of randomness in sampling. If it is so important why are sampling methods used which do not include random methods?

3 Find a description of a sample survey and summarise in no more than 500 words the main sampling method used.

4 Design a sample survey to discover the likely success of a new consumer durable, carry out a pilot survey (consider possible products and names).

5 What are the objectives of sampling? Consider why sample surveys are carried out on preference to full surveys.

6 Carry out a series of dice-throwing exercises consisting of a total of 100 throws. Record the numbers that come up on each throw and the numbers that come up in each set of 10 throws, in the form of a table. Draw a curve to illustrate the table. Write a report to indicate the effects of changing sample size and to compare the results against the expected frequencies.

CHAPTER 6
HOW TO USE FIGURES

6.1 THE NEED TO USE FIGURES

Once information has been collected by a full survey or a sample, it has to be put into a form which can be used easily. The raw survey data has to be processed, analysed, summarised and presented. Much of this process is descriptive in the form of reports, tables, graphs and averages but also inferences can be drawn from the data based on its accuracy.

Both the descriptive and inferential aspects of statistics require the use of basic mathematics. The objectives of this chapter are to provide a reminder of the main areas of mathematics necessary to understand statistics.

6.2 THE VOCABULARY OF MATHEMATICS

It can be argued that the main purpose of a language is to enable people to exchange ideas with the minimum of effort and the maximum of clarity. Mathematics has its own language, with a vocabulary of its own in the form of numbers and symbols (see Section 6.16). The basic vocabulary of mathematics is a number system:

(a) The decimal system
This is based on groups of ten (from the Latin *decem*, meaning ten). The system was invented by the Hindus about 1500 years ago and was passed on by the Arabs towards the end of the eleventh century.

The system is based on the idea of positional value and a base of ten. All numerals are constructed with ten basic symbols, and in writing numerals the actual position of any given numeral is significant.

The basic symbols are 0 to 9. To represent numbers ten times as large, these digits are shifted one position to the left, the digit 0 being used as a position indicator: 10, 20, 30 . . . 90. Further increases by a factor of ten are indicated by further shifts in position: 100, 200, 300. . . .

A decrease by a factor of ten is shown by shifting the digits one position to the right, the digit 0 again being used when necessary to indicate position: one-tenth is 0.1, one-hundredth is 0.01, one-thousandth is 0.001.

For example:

Thousands	Hundreds	Tens	Units
1	4	6	2
5	9	4	1
	7	3	2
2	0	3	6
10	1	7	1

The positions of the numbers in the columns are important; a number in any column represents ten times the same number in the column on its right and one-tenth of the same number in the column on its left.

(b) **The binary system**

This has a base of two (and is used in computing). The two digits used are 0 and 1 and any number can be represented by locating these digits in appropriate positions. Zero is 0, one is 1, two is 10, three is 11, four is 100, five is 101, six is 110, seven is 111, eight is 1000.

An increase in a number by a factor of two (that is, doubling it) is shown by shifting the digits one position to the left, 0 being used to indicate position. A decrease by a factor of two (halving the number) is shown by shifting the digits one position to the right.

For example: five is 101, ten is 1010, twelve is 1100, six is 110.

Systems based on numbers other than ten or two are possible, but these two systems are the most common.

6.3 BASIC ARITHMETIC

The four basic arithmetical operations are addition, subtraction, multiplication and division, with the symbols +, −, ×, ÷ respectively.

(a) **Addition** is the process of putting numbers together. The sign + (or *plus* from the Latin for 'more') means that the number following it is to be added to the number preceding it. The answer obtained by adding numbers together is called the 'sum' (with the symbol Σ or sigma), and the answer is indicated by the sign = (equals).

Positive numbers are normally written without a plus sign, while with negative numbers the minus is always written. The value of a number without its sign is called its 'absolute' value. With −2 and +2, the number 2 is the absolute value of both.

(b) **Subtraction** is the process of finding the difference between two numbers. It is the inverse of addition. The sign – (or *minus*, from the Latin for 'less') indicates that the number following is to be taken away from the number preceding it.

(c) **Multiplication** is the process of finding the sum of a number of quantities which are all equal to one another. It is essentially a short-cut version of adding when all the numbers are the same. Therefore 8×6 means $8 + 8 + 8 + 8 + 8 + 8 = 48$. Multiplication is indicated by the sign × (or *times*) and the result of the multiplication is called the 'product'.

When two negative numbers are multiplied together they make a positive product, while a positive and negative number multiplied produces a negative product ($-8 \times -6 = 48$, but $8 \times -6 = -48$).

(d) **Division** is the process of finding out how many times one number is contained in another number. It is the inverse of multiplication. Division is indicated by the sign ÷ (or *divided by*).

6.4 THE SEQUENCE OF OPERATIONS

Addition, subtraction, multiplication and division must be carried out in the correct order in mathematical calculations. The mnemonic *BODMAS* summarises this sequence:

>*B*rackets
>*O*f
>*D*ivision
>*M*ultiplication
>*A*ddition
>*S*ubtraction

This could be pronounced *B – ODM – AS*, because *ODM* have equal priority and *AS* have equal priority. 'Of' stands for multiplication, as in "$\frac{1}{2}$ of 10".

For example:

>(i) $5 + 2 \times 3 = 11$ *not* 21 ($2 \times 3 = 6 + 5 = 11$, *not* $5 + 2 = 7 \times 3 = 21$)
>(ii) $(5 + 2) \times 3 = 7 \times 3 = 21$

This example illustrates the fact that:

(i) Brackets should be evaluated first.
(ii) If an expression contains only pluses and minuses or only multiplication and division, then the sum should be calculated by working from the left to the right.
(iii) Multiplication and division should be calculated before addition and subtraction.

6.5 SIMPLE ARITHMETIC

The following sums are included to illustrate the rules of simple arithmetic. If there are any doubts about these rules, then ideally the sums are worked out before looking at the answers and explanations.

(i) 30 − 9 + 3: In this case the working is from left to right (24).
(ii) 8 × 3 − 2: The Multiplication comes before the subtraction (22).
(iii) − 5 × 3: A minus times a plus equals a minus (−15).
(iv) − 5 × − 3: Two minuses make a plus, in the same way that a double negative cancels out: *'not im*possible' means that it is possible (15).
(v) 4 + 7 × 3: Multiplication comes before addition (25).
(vi) 4(3): A bracket indicates multiplication. This is another way of writing 4 × 3 (12).
(vii) 4(3 + 1): The bracket always comes first (16).
(viii) (2 − 1) (7 + 5): The brackets come first, the two brackets indicate multiplication (12).
(ix) 5 − 3 × 4 + 5: Multiplication comes before subtraction and addition (−2).
(x) 122 × 28 ÷ 7 + 60: Multiplication and division come first (but have equal priority), addition last (548).

6.6 FRACTIONS

Fractions allow the consideration of units of measurement smaller than a whole number. The term 'common fraction' is used to emphasise the distinction from the decimal fraction: $\frac{1}{2}$ as opposed to 0.5.

$$\text{Common Fraction} = \frac{\text{Numerator}}{\text{Denominator}}$$

A 'proper fraction' is one where the numerator is less than the denominator ($\frac{1}{2}$ or $\frac{3}{4}$). An 'improper fraction' is where the numerator is greater than the denominator: $\frac{53}{10}$. This can be reduced to a whole number and a proper fraction: $5\frac{3}{10}$.

(a) The addition and subtraction of fractions

Fractions often occur when things are measured or when they are divided. The classic example is dividing a cake into slices. Assume that the cake is divided equally between three people; it will be divided into three equal slices or thirds:

$$\frac{1 \text{ whole cake}}{3 \text{ people}} = 3 \text{ slices, each } \tfrac{1}{3} \text{ of a whole cake.}$$

Fig 6.1 *slices of cake*

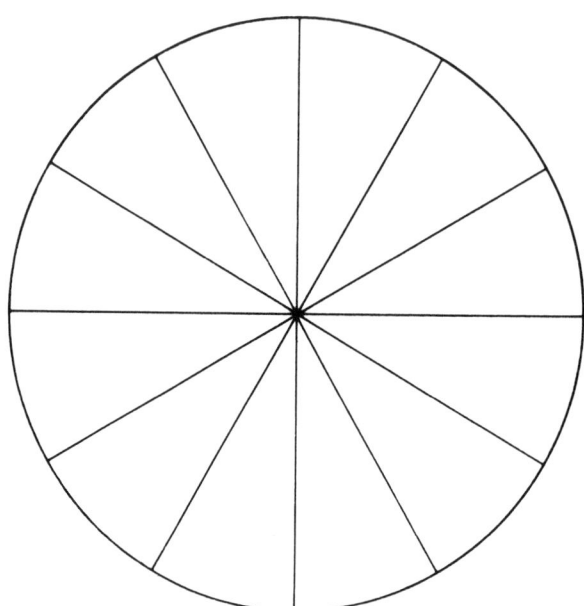

If one person eats $\frac{1}{3}$ of one cake and $\frac{1}{4}$ of another cake, in total he has eaten $\frac{1}{3} + \frac{1}{4}$ pieces of cake. To add these the common denominator is found, that is the number of slices into which both $\frac{1}{3}$ and $\frac{1}{4}$ can be divided with whole numbers. If a cake is divided into twelve slices, it is possible to count how many slices of cake the person has consumed; $\frac{1}{3}$ = 4 slices and $\frac{1}{4}$ = 3 slices. Therefore $\frac{1}{4} + \frac{1}{3} = \frac{3}{12} + \frac{4}{12} = \frac{3+4}{12} = \frac{7}{12}$.

If the person was now told that $\frac{5}{12}$ths of the cake would be taken away to give to other people, then he would have $\frac{7}{12} - \frac{5}{12} = \frac{2}{12}$ or $\frac{1}{6}$th of the cake left. This can be checked in Figure 6.1.

(b) The multiplication and division of fractions

To multiply fractions the numerators are multiplied together to obtain the numerator of the answer, and the denominators are multiplied together to obtain the denominator of the answer.

For example: $\frac{2}{3} \times \frac{4}{12} = \frac{6}{36} = \frac{1}{6}$

To divide by a fraction, the fraction is multiplied by its inverse.

For example: $\frac{4}{7} \div \frac{5}{8} = \frac{4}{7} \times \frac{8}{5} = \frac{32}{35}$

Another example: $125 \div \frac{1}{5} = 125 \times \frac{5}{1} = 625$

To divide mixed fractions they must first be made improper (a 'mixed fraction' is one that contains whole numbers as well as fractions).

For example:

$$3\tfrac{2}{5} = 3 + \tfrac{2}{5} = \tfrac{15}{5} + \tfrac{2}{5} = \tfrac{17}{5}$$

$\therefore \quad 3\tfrac{2}{5} \div 2\tfrac{4}{7} = \tfrac{17}{5} \div \tfrac{18}{7} = 1\tfrac{29}{90}$

6.7 DECIMALS

A decimal number is one whose denominator is 10, or 100, or 1000, or any power of ten. For instance, the decimal number 'three-tenths' is written $\tfrac{3}{10}$ or 0.3. The decimal point divides the whole number from the fraction. 1.3 equals the whole number 1, plus the fraction 0.3. The decimal point should not be confused with a full stop. The decimal point can be written above the line, as in 0·3, to avoid this ambiguity.

When adding and subtracting numbers which include decimals, the decimal points must be kept underneath one another to avoid the difficulties of place value.

For example: add up 1.16, 2.75, 0.08, 8.057

```
 1.16
 2.75
 0.08
 8.057
_____
12.047
```

Applications of the decimal system include money and the metric system:

(a) Decimal currency

The decimal currency of the UK is based on the pound sterling (£) and is divided into 100 pence (p). Sums of less than £1 in value can be written in two ways:

(i) 55p
(ii) £0.55

If the number of pence is less than 10, there should be a 0 in the ten pence column to indicate that there are not any ten pence pieces involved:

£0.08 = 8p   £0.80 = 80p

(b) The metric system

There is a general world trend towards using the metric system of weights and measures. The name is derived from the basic unit of length, the metre.

The International System of Units (referred to as SI units) covers all types of measurement:

> the metre (m) for length
> the litre (l) for volume and capacity
> the kilogram (kg) for weight
> the degree Centigrade (°C) for temperature
> the second (s) for time

The tables for all these measures are based on units of ten:

For example:

Table of length
10 millimetres (mm) = 1 centimetre (cm)
100 centimetres (cm) = 1 metre (m)
1000 metres (m) = 1 kilometre (km)

Table of capacity
1000 millilitres (ml) = 1 litre (l)
1000 litres (l) = 1 kilolitre (kl)

Table of weights
100 milligrams (mg) = 1 gram (g)
1000 grams (g) = 1 kilogram (kg)
1000 kilograms (kg) = 1 tonne or metric tonne (t)

(c) The imperial system

The metric system is gradually replacing the imperial system of weights and measures, but the imperial system is still used and it is likely to be of both practical interest and historical interest for some time. The imperial system is based on:

> the foot (f) for length
> the pint (p) for volume and capacity
> the pound (lb) for weight
> the degree Fahrenheit (°F) for temperature
> the second (s) or minute (m) for time.

The tables for all of these are based on a variety of units of numbers;

For example:

Table of length
12 inches = 1 foot

3 feet = 1 yard
1760 yards = 1 mile
(1 metre = 39.37 inches or 3 feet 3.37 inches)

Table of capacity
2 pints = 1 quart
4 quarts or 8 pints = 1 gallon
(1 litre = 1.76 pints)

Table of weight
16 ounces (oz) = 1 pound (lb)
14 pounds (lb) = 1 stone
112 pounds (lb) = 1 hundredweight (cwt)
20 hundredweight (cwt) = 1 ton
(1 kilogram = 2.2 pounds (lb))

Table of temperature
Freezing in Celsius is at 0°, in Fahrenheit at 32°. Conversion formulas are:

$$C = \tfrac{5}{9}(F - 32) \text{ and } F = \tfrac{9}{5}C + 32$$

Therefore if F = 60° then:

$$C = \tfrac{5}{9}(60 - 32)$$
$$= 15.55\,°C$$

If C = 15° then:

$$F = \tfrac{9}{5} \times 15 + 32$$
$$= 15.55\,°C$$

6.8 PERCENTAGES

'Per cent' (or percent) means per hundred (from *centum*, the Latin for hundred). Therefore 50 per cent is 50 per hundred or 50 out of a hundred, that is, one half. The sign for a percentage is % which contains the one, zero, zero of 100.

(i) To change a fraction to a percentage multiply by 100: $\tfrac{1}{4}$ equals 25% ($25 = \tfrac{1}{4} \times 100$).
(ii) To change a decimal to a percentage multiply the decimal by 100 and put the % after the result: $0.55 \times 100 = 55\%$.
(iii) To change a percentage to a fraction, divide by 100: $50\% = \tfrac{50}{100} = \tfrac{1}{2}$.
(iv) To change a percentage to a decimal, divide by 100 by moving the decimal point two places to the left: $13.96\% = 0.1396$.
(v) Finding a percentage can be seen by example:

£7 as a % of £50 = $\tfrac{7}{50} \times 100 = \tfrac{700}{50} = 14\%$

5% of £4.50 = $\tfrac{5}{100} \times £4.50 = \tfrac{£22.50}{100} = 22.5\text{p}$

6.9 POWERS AND ROOTS

Multiplication is a shorthand way of representing a series of additions; a shorthand method of representing a series of multiplications is by the use of 'powers'.

For example: $4 \times 4 \times 4$ is four raised to the power of three, or $4^3 = 64$ ($4 \times 4 = 16 \times 4 = 64$).

Four power two (4^2) is four squared.

The square root of a number is that number which when multiplied by itself gives the original number. Square root is shown by the sign $\sqrt{}$

For example: the square root of 16 is 4, because $4 \times 4 = 16$.

Therefore $\sqrt{16} = 4$

The square roots of many numbers are difficult to calculate exactly, although they can be estimated through trial and error.

For example: the square root of 30 lies between 5 and 6 ($5 \times 5 = 25$, $6 \times 6 = 36$). A close estimate would be 5.5. Correct to two decimal places the square root of 30 is 5.48 ($5.48 \times 5.48 = 30.02$).

Many calculators do not have a square-root key and therefore to reduce the time taken by trial and error, square-root tables are provided at the end of this book (Appendix A.3).

Two sets of tables are provided. Tables A and B for numbers from 1 to 10, and Tables C and D for numbers from 10 to 100. The square root of 1 is 1, and the square root of 100 is 10, but the square root of 10 is 3.1623 and the square root of 1000 is 31.623. Therefore if the square root of a number is multiplied by ten, the number by which the new answer is a square root would be a hundred times larger. So that only two sets of tables are needed to obtain the square root of any number.

To obtain the square root of a number the following procedure can be followed:

(i) Make a very rough estimate of the size of the square root:

For example: $\sqrt{3.204}$

1^2 is 1, and 2^2 is 4, so the answer is between 1 and 2.

$$\sqrt{320.4}$$

15^2 is 225, and 20^2 is 400, so the answer is between 15 and 20.

$$\sqrt{3204}$$

50^2 is 2,500, and 60 is 3,600, so the answer is between 50 and 60.

(ii) If the number whose square root is being obtained is between 1 and 100, look up the nearest figure on the table:

For example: $\sqrt{3.204} = 1.790$

If it is not between 1 and 100, multiply or divide it successively by 100 until it does lie between 1 and 100. Then look up the square root of the result, and multiply or divide it successively by 10.

For example: $\sqrt{0.3204}$: $0.3204 \times 100 = 32.04$

$\sqrt{32} = 5.657$ $5.657 \div 10 = \underline{0.5657}$

$\sqrt{3.204}$: $3.204 \div 100 = 32.04$

$\sqrt{32} = 5.657$ $5.657 \times 10 = \underline{56.57}$

(iii) Check that the answer is within the limits estimated in stage (i).

6.10 RATIOS

A ratio is a relationship between two quantities expressed in a number of units which enables comparison to be made between them.

For example:

(i) Two motor-cars may be travelling at different speeds, say 60 km per hour and 30 km per hour. The ratio of speeds to one another is said to be: 60 : 30, or 6 : 3, or 2 : 1.

(ii) Three-quarters of the annual output of a factory may consist of commodity A and one-quarter of commodity B. The ratio of output can be said to be 3 : 1. For every 3 units of A produced in a year, 1 unit of B is produced.

6.11 PROPORTIONS

Many business calculations are based on simple proportions.

For example:

(i) If 5 kilograms of copper cost £24 what will 12 kilograms of copper cost? If the price is the same for every kilogram, then the higher cost of the 12 kilograms will be in direct proportion:

5 kg = £24
1 kg = £4.8 ($£\frac{24}{5}$)
12 kg = £57.6 (£4.8 × 12)

(ii) If a stay at a hotel costs £112 for 7 days, how much will it cost for 15 days at the same rate?

7 days = £112
1 day = £16 ($£\frac{112}{7}$)
15 days = £240 (£16 × 15)

6.12 ELEMENTARY ALGEBRA

Algebra provides a method of abbreviating information without loss of clarity or accuracy:

For example: an abbreviation of a basket of fruit could be written as: 10 As, 5 Os, 8 Bs, where A = apples, O = oranges and B = bananas. Equally these three types of fruit could be labelled x, y and z. If another basket of fruit was said to contain $15x$ and $10y$, this could be translated as 15 apples and 10 oranges.

The expression $15x$ means 15 times x. Therefore if x is also a number, say 3, then the value of $15x$ when $x = 3$ is 45.

This should not be confused with the use of letters such as x and y as symbols in statistical calculations, where they often denote two variables.

6.13 LEVELS OF MEASUREMENT

In mathematics there are different levels of measurement, from a basic labelling system to a system which indicates the value of one number against another.

(a) Nominal scales
This is the simplest level of measurement. It is a method of classification and the function of numbers is the same as names when labelling categories. There is no implication that one is better or greater than another.

For example: hotel rooms have numbers to distinguish them, but usually not to indicate which is the best or worst room. There is no point in adding up hotel room numbers; it may be useful to know that there are ten rooms but it is not likely to be useful to know that the numbers add up to 55.

Nominal scales possess the properties of symmetry and transitivity. Symmetry means that a relationship between A and B also is true between B and A. If A is opposite to B, then B is opposite to A.

Transitivity means that if $A = B$ and $B = C$, then $A = C$. If A is the same age as B, and B is the same age as C, then A must be the same age as C.

(b) Ordinal scales
This is the ordering of categories with respect to the degree to which they possess a particular characteristics, without being able to say exactly how much of the characteristic they possess.

For example: workers can be classified into 'unskilled', 'semi-skilled' and 'skilled' without giving the exact interval between these classifications.

Another example: three factories may be put into order by size: 1, 2 and 3. This will not give any indication of relative sizes. In fact the factories may employ 10,000 people, 8000 people and 30 people respectively, but these differences are not shown on the ordinal scale.

Ordinal scales are asymmetrical. This means that a special relationship may hold between A and B which does not hold between B and C. If A is greater than B, then B cannot be greater than A.

Ordinal scales do have the property of transitivity. If A is greater than B and B is greater than C, then A will be greater than C.

(c) Interval scales

These scales not only rank objects with respect to the degree with which they possess a certain characteristic, but also indicate the exact distances between them. This requires a physical unit of measurement which can be agreed upon as a common standard and that can be applied over and over again with the same results.

Examples of these units of measurement are the metric system and the imperial system. Length, for example, is measured in metres or feet. There are no such units of intelligence or authoritarianism which can be agreed upon.

Given a unit of measurement, it is possible to say that the difference between scores is twenty units, or that one difference is twice as large as a second. Scores can be added and subtracted and so on.

(d) The scales compared

An ordinal scale possesses all the properties of a nominal scale plus ordinality. An interval scale has all the properties of a nominal and ordinal scale plus a unit of measurement.

The cumulative nature of these scales means that it is always possible to drop back one or more levels of measurement in analysing data. Sometimes this is necessary when statistical techniques are unavailable or unsatisfactory for handling the variable on an interval scale. However, by using a lower scale information is lost:

For example: if it is known that A has an annual income of £15,000, B has an income of £14,000 and C has an income of £5000, it is possible to use the ordinal scale and say that A has the highest income, B the second highest and C the lowest. In order of size of incomes the ranking is $A : 1$, $B : 2, C : 3$.

However, using the ordinal scale in this example throws away the information that the difference in incomes is small between A and B and much larger between A/B and C, and that A has an income that is three times as great as that of C. Therefore it is advantageous to make use of the highest level of measurement that can be used.

6.14 FINANCIAL MATHEMATICS

Certain aspects of mathematics have a very direct connection with finance; an important example of this is the link between the arithmetic and geometric progressions and simple and compound interest.

(a) Simple interest
An arithmetic progression is where in a series of numbers the difference between them is the same.

For example:

(i) 3, 9, 12, 15, 18, 21 (the difference is 3)
(ii) $1\frac{1}{2}$, 3, $4\frac{1}{2}$, 6, $7\frac{1}{2}$ (the difference is $1\frac{1}{2}$)

This concept is used for calculations of simple interest.

For example: if £100 is invested for 4 years at a simple interest rate of 5% per annum, at the end of 4 years the total amount accumulated would be:

$$£100 + £5 + £5 + £5 + £5 = £120$$

The total amount was grown in arithmetic progression: £100, £105, £110, £115, £120.

This can be shown by a formula:

$$A = P(1 + tr)$$

where A is the total amount accumulated
P is the original investment
t is the time in years
r is the rate of interest

Therefore: $A = £100 (1 + 4 + \frac{5}{100})$

$= £120$

(b) Compound interest
A geometric progression is where in a series of numbers the difference between the numbers is found by multiplying the preceding number by a fixed amount (often called the 'common ratio').

For example: 4, 8, 16, 32, 64 (each number is multiplied by 2 to arrive at the following number).

This concept is used in calculations of compound interest. It is often the case, when a sum of money has been invested, for the interest payment (in a bank account or building society) to be reinvested rather than withdrawn and spent. This is the basis of compound interest.

For example: if £100 is invested for 4 years at a compound interest rate of 5% per annum, at the end of 4 years the total amount accumulated would be:

original investment = £100
Year 1 = £105
Year 2 = £110.25
Year 3 = £115.76
Year 4 = £121.55

This can be shown by a formula:

$A = P(1 + r)^t$

where A is the total amount accumulated
P is the original investment
r is the rate of interest
$(1 + r)$ is the 'common ratio'
t is the time in years

The original investment is multiplied by the common ratio to the power of t, that is the number of years which the investment lasts.

Therefore $A = £100 (1 + \frac{5}{100})^4$

$= 121.55$

Notice that P is known as the 'present value' of A at a compound interest rate (r) in t years from now. So that £100 is the present value of £121.60 at a 5% rate of interest in 4 years time.

Money invested at compound interest quickly builds up to a large sum: £100 at 5% over 10 years would produce an accumulation of £162.89 while at simple interest this would be £150. This is why compound interest has become so important. Most savings such as life assurance policies and pension funds are partly built up from the compound interest they earn.

(c) **Present values**

The concept of present value works in the opposite direction to the calculation of compound interest. The kind of question asked is: 'What sum of money, if invested at an interest rate of 10% per annum compounded annually, will give £100 in five years time?"

The formula is:

$$P = \frac{A}{(1+r)^t}$$

where P is the original investment or the present value

A is the total amount accumulated
r is the rate of interest
t is the time in years

Therefore

$$P = \frac{100}{(1 + \frac{10}{100})^5}$$

$$= £62$$

Thus £62 is the present value (or discounted value) of £100 due at the end of five years; £62 now is equivalent to £100 in five years time. This has nothing to do with inflation. It can be argued that £1 now is always worth more than £1 in the future, simply because the £1 can be used to earn more money. When a business borrows money, it is in effect exchanging a larger sum of future money for a smaller sum of present money, which it can use profitably.

When calculating present values, the discount rate used is normally the prevailing rate of interest on money borrowed.

Present values have a number of financial applications. One example is *discounted cash flow* which involves the calculation of the present value of a series of future cash flows. Cash flow is the difference between the money flowing into a business (from receipts) over a certain period and the money flowing out (from payments).

By calculating present values or the net present value of a proposed investment project a business can ascertain its present worth. Faced with a choice between alternative investments, businesses can calculate their net present values to provide a valid means of comparison.

For example: a firm has a choice between two plants, one of which costs £50,000, the other £40,000. They each have a useful life of four years. The problem is which plant the firm should purchase if the discount rate is 20%.

	Cost	Estimated Annual Cash Flow (£)			
Plant A	−£50,000	+10,000	+20,000	+40,000	+30,000
Plant B	−£40,000	+10,000	+20,000	+15,000	+ 5,000

Plant A's net present value (in £000s):

$$-50 + \frac{10}{1.2} + \frac{20}{1.2^2} + \frac{40}{1.2^3} + \frac{30}{1.2^4} = £9830$$

Plant B's net present value (in £000s):

$$-40 + \frac{10}{1.2} + \frac{20}{1.2^2} + \frac{15}{1.2^3} + \frac{5}{1.2^4} = £-6690$$

The cash flow for each year is divided by 1 plus the rate of interest (20% or 0.2) to the power of the time in years (as in the formula for present values above). Plant A has the higher net present value. The negative net present value of Plant B shows that this plant will yield a return on investment which is less than the current discount rate. A project with a negative net present value is not viable since either the firm will not be able to cover the cost of borrowing, or if it has spare cash, it can invest it more profitably elsewhere.

(d) **Discounted rate of return**

This is concerned with the discount rate which will give a net present value of zero.

In the previous example, if the firm has to buy plant B or give up the project, the problem is to find the discount rate that will give a zero net present value and therefore make the project viable. The 20% discount rate applied in the example gives a negative net present value. For a zero net present value, the discount rate must be less than 20%. To determine what this is it is necessary to proceed by trial and error.

A discount rate of 10% gives a net present value of £ + 0.3200. This is just positive. A discount rate of 11% gives a net present value of £–0.500. This is just negative. Therefore the expected returns from this project will repay the original investment of £40,000 if the firm can borrow the money at an interest rate of 10% or less.

(e) **Decisions**

These techniques may help a firm to make a decision on investment, loans and borrowing. In fact these decisions are complicated by taxation, government grants and so on. These can be appropriately discounted.

Many investment decisions are made on the basis of other factors, other costs and benefits which may include such factors as the environment, industrial relations, 'hunches' and so on which are not easily 'discounted'.

6.15 AIDS TO CALCULATION

(a) **Electronic calculators**

These first came into common use in the early 1960s. Over the years their price has fallen considerably and they are increasingly widely used.

Calculators are of many different kinds, but a relatively inexpensive pocket calculator will perform all the calculations needed in an ordinary business. All calculators will add, subtract, multiply and divide. For business purposes it may be useful for a calculator to have other attributes:

(i) For frequent calculations of such things as Value Added Tax, a constant factor facility is useful. This allows the multiplication or

division of a series of numbers by the same number without having to enter each calculation separately.
(ii) Some models calculate percentages, mark-ups and discounts directly.
(iii) A memory facility may be useful for complicated calculations.
(iv) A clear last entry facility allows a mistake to be corrected in an entry without having to go all the way back to the beginning of the calculation.

(b) Logarithms

For 250 years before electronic calculators came into everyday use, people used logarithms to help them carry out laborious calculations. The word comes from the Greek for 'calculating with numbers'.

The use of logarithms avoids long multiplication and division calculations, substituting the relatively simpler addition and subtraction processes.

Logarithms are exponents and are based on the same principles as exponents. Exponents provide a shorthand method of writing out multiple multiplication, they are the symbol showing the power of a factor.

For example: $2^6 = 2 \times 2 \times 2 \times 2 \times 2 \times 2 = 64$
6 is the exponent providing the instruction to multiply 2 by itself 6 times.

In common logarithms the logarithm of a number is the exponent of 10 that will produce that number.

For example:

$10^0 = 1$ $\qquad 10^{-1} = \frac{1}{10} = 0.1$
$10^1 = 10$ $\qquad 10^{-2} = \frac{1}{100} = 0.01$
$10^2 = 100$ $\qquad 10^{-3} = \frac{1}{1000} = 0.001$

Therefore the log of:

1	= 0	0.1	= −1
10	= 1	0.01	= −2
100	= 2	0.001	= −3
1000	= 3	0.0001	= −4

Therefore the logarithm of all numbers
between   1   and   10 will be 0 plus a fraction
           10   and   100 will be 1 plus a fraction
          100   and   1000 will be 2 plus a fraction
            1   and   0.1 will be −1 plus a fraction
         0.1   and   0.01 will be −2 plus a fraction
     0.01   and   0.001 will be −3 plus a fraction

The logarithm of a number consists of two parts, the integer or characteristic and the fractional part or mantissa.

For example:

$$\text{Log } 70 = 1.8451$$

Characteristic   Mantissa
(Integer)   (Fraction)

(i) *The characteristic*: when a number is greater than one the characteristic of the logarithm is positive; if the number is less than one, the characteristic is negative and is called 'bar one', 'bar two' and so on with the minus sign written over the characteristic (e.g. $\bar{1}$).

For example:

$$\text{Log } 70 = 1.8451$$
$$0.7 = \bar{1}.8451$$

The characteristic of the logarithm of a number is one less than the number of integers it contains.

For example:

Log	Characteristic
7	0
70	1
700	2

The characteristic of a logarithm of a decimal or fraction is negative and is equal to the number of places occupied by the first significant figure of the decimal.

For example:

Log	Characteristic
0.7	$\bar{1}$
0.07	$\bar{2}$
0.007	$\bar{3}$

(ii) *The mantissa*: this decimal part of a logarithm is obtained from tables, 'logarithm tables' or 'four-figure logarithm tables' (see Appendix A.4). The term 'four-figure' tables means that the mantissa is given correct to four decimal places. The characteristic is obtained by inspection and is omitted from the tables because its function is to indicate the placing of the decimal point in the result.

The mantissa is the same for all numbers made up of the same series of digits. The position of the decimal point changes the characteristic but not the mantissa.

For example:

Log 70.53 = 1.8484

The characteristic is obtained by inspection. The mantissa is found by looking up the first two figures (70) in the first left-hand column of the logarithm tables. Then the column headed by the third figure (5) is consulted. Where the row 70 meets the column 5 is the number 8482. The fourth number (3) is looked up in the columns on the right-hand side of the tables (if there are more than four numbers, then the fourth number is rounded). The column headed 3 meets the row 70 at a number 2. This is added to 8482 to arrive at the mantissa for the numbers 7053 (8482 + 2 = 8484).

Therefore the log of:

$$705.3 = 2.8484$$
$$70.53 = 1.8484$$
$$7.053 = 0.8484$$
$$0.7053 = \overline{1}.8484$$

Extract from four-figure logarithm tables

	0	1	2	3	4	5	6	7	8	9	1	2	3	4	5	6	7	8	9
70	8451	8457	8463	8470	8476	8482	8488	8494	8500	8506	1	1	2	2	3	4	4	5	6
71	8513	8519	8525	8531	8537	8543	8549	8555	8561	8567	1	1	2	2	3	4	4	5	5

(iii) *The antilogarithm*: is used to return the log number to the original number. If the log number is 1.8484, the row beginning .84 is looked up (in 'antilogarithm tables') and the column headed 8. Where this row and column meet is the figure 7047. The column headed 4 on the extreme right meets the row 84 at the number 6. When this is added to 7047 the result is the figure 7053. Since the characteristic of 1.8484 is the number 1, then two figures are added to the left of the decimal point. Therefore the antilog of 1.8484 is 70.53. The antilog of 2.8484 is 705.3, the antilog of 3.8484 is 7,053 and so on.

(iv) *Use of logarithms*: they are especially useful in performing calculations involving multiplication and division and finding the power and root of a number.

(a) Multiplication: to multiply two numbers their logarithms are added.

For example: 70.53 × 7.126

$$\begin{aligned} \text{Log } 70.53 &= 1.8484 \\ \text{Log } 7.126 &= 0.8529 \\ \hline &\,2.7013 \end{aligned}$$

Antilog 2.7013 = 502.7

Therefore 7053 × 7.126 = 502.7

(b) Division: to divide one number by another, the logarithm of the denominator is subtracted from the logarithm of the numerator.

For example: 70.53 ÷ 7.126

$$\begin{aligned}\text{Log } 70.53 &= 1.8484\\ \text{Log } 7.126 &= 0.8529\\ \hline &\,0.9955\end{aligned}$$

Antilog 0.9955 = 9.897

Therefore 70.53 ÷ 7.126 = 9.897

(c) Power: to find the power of a number the logarithm is multiplied by the required power.

For example: 70.53^3

$$\begin{aligned}\text{Log } 70.53 &= 1.8484\\ &\times 3\\ \hline &\,5.5452\end{aligned}$$

Antilog 5.5452 = 351,000

Therefore 70.53^3 = 351,000

(d) Root: to find the root of a number the logarithm of the number is divided by the required root (see also Section 5.9).

For example: the square root of 70.85 ($\sqrt{70.53}$)

Log 70.53 = 1.8484

1.8484 ÷ 2 = 0.9242

Antilog 0.9242 = 8.399

Therefore $\sqrt{70.53}$ = 8.399

(e) Log scales (or ratio scales): logarithms are essential for marking out log scales. These are used in semi-log graphs where the *y* or vertical axis is measured in a logarithmic or ratio scale (see Section 8.9).

6.16 SYMBOLS OF MATHEMATICS

This list includes the main mathematical symbols used in statistics:

> x is a collective symbol meaning all the individual values of a variable.
> y is an alternative symbol to x. It is used where there are two sets of variables and x has already been used to indicate the first.
> $\bar{x}$ (bar x or x bar): a bar over a variable symbol indicates that it represents the arithmetic mean of the values of that variable.
> n stands for the number of items in a collection of figures (10 apples, n 10).
> f stands for *frequency*; that is the number of times a given value occurs in a collection of figures.
> Σ means the *sum of* (Σ is the Greek capital S or sigma).
> σ (small sigma) stands for the *standard deviation*.
> d stands for *deviation*, which is the difference between two values.
> r stands for the *coefficient of correlation*.
> r' stands for the *coefficient of rank correlation*.
> $=$ means *equals*.
> $\simeq$ means *approximately equals*.
> $\neq$ means *not equal to*.
> $>$ means *greater than* or *larger than* or *more than*.
> $<$ means *smaller than* or *less than*.

ASSIGNMENTS

1   Calculate the following:

(i) $6 + 4 \times 2$
(ii) $-2(2 - 1)$
(iii) $\frac{1}{4} - \frac{1}{8}$
(iv) $6\frac{3}{4} - 3\frac{7}{8}$
(v) $0.92 - 0.42$
(vi) $9 \div 0.03$
(vii) £6 as a percentage of 510
(viii) 20% of £5
(ix) evaluate 2^7
(x) evaluate $\sqrt{0.01}$
(xi) evaluate $(1.4)^2$
(xii) what is $5x + 6y$ when $x = 3$ and $y = 4$?

2   In an office of 640 employees $\frac{3}{8}$ are women and $\frac{5}{8}$ are men. How many (a) women and (b) men are there?

3 What sum of money, if invested at an interest rate of 20% per annum compounded annually, will give £500 in 3 years' time?

4 Discuss why the highest possible level of measurement should be used.

5 Using logarithm tables calculate the following:

(i) 714.8×703.6
(ii) $714.8 \div 7.132$
(iii) $\sqrt{712.6}$

6 Find examples of the use of compound interest in the economy. Write a report considering its use in these cases.

CHAPTER 7
HOW TO PRESENT FIGURES (1)

7.1 THE AIMS OF PRESENTATION

Primary and secondary data need to be arranged and presented in some way before the information contained in the data can be interpreted. Primary data may be in the form of a pile of completed questionnaires or a long list of figures; secondary data may be contained in government publications, company reports, books and archives from which relevant information needs to be collated and presented in a form which can be appreciated.

The aim of presenting figures is to communicate information. Therefore the type of presentation will depend on the requirements and interests of the people receiving the information.

For example: to compare different methods of presenting the same information, all the national newspapers can be bought on the day after a Budget. It is then possible to compare the ways in which the figures contained in the Budget are presented. The 'posh' papers tend to print tables of figures and an almost verbatim report of the Budget. The tabloids publish diagrams and drawings and brief tables, with a short summary of the Budget speech in Paliament.

When data is presented it is important to provide information clearly and at the same time make an impact. The sales graph which shows a sharp decline in sales from the date the new sales manager arrived, is a good example (Figure 7.1):

It has been suggested that the facts do not matter as much as the way they are presented. Obviously facts do matter, but if they are poorly presented they may be overlooked. Therefore the way in which statistical data is presented is important.

The form of the presentation should be based on the following factors:

Fig 7.1 *sales graph for Company* A

(i) A clear presentation of the subject matter.
(ii) A clarification of the most important points in the data.
(iii) Consideration of the purpose of the presentation.
(iv) Consideration of the amount of detail and accuracy required.
(v) The use of the most appropriate method of presentation.

For example: presenting sales figures to a meeting of experts may require tables, a report, graphs and diagrams, showing great detail. A group of shareholders may be interested in seeing only a graph which shows whether sales are rising or falling.

Frequently the first stage in presenting figures is to produce a table. Often this will be constructed before a report is written or graphs and diagrams drawn, because these will be based on the information contained in the table.

7.2 TABULATION

(a) The purpose of tabulation
Tables are used to facilitate the understanding of complex numerical data. Therefore a table should be as simple and unambiguous as possible.

There are a number of types of table, examples of which can be seen in government publications (such as *Financial Statistics*, the *Annual Abstract of Statistics*, *Economic Trends*, *Social Trends*), journals and magazines (such as the bank reviews and *The Economist*) and in company reports. Tables can be divided into:

(i) Informative tables providing a statistical record.
(ii) Reference tables containing summarised information backed up by more complex tables providing a complete analysis.
(iii) Complex tables showing a number of columns and rows which are interrelated with sub-totals and totals divided into a number of cateogries.

Tables are used:

(i) To present the original figures in an orderly manner.
(ii) To show a distinct pattern in the figures.
(iii) To summarise the figures.
(iv) To provide information which may help to solve problems.

(b) **The construction of tables**

In the construction of statistical tables there are guidelines which it is important to consider:

(i) All tables should have a title which gives an indication of the contents.
(ii) The source of the data must be included (usually below the table) so that the original sources can be checked.
(iii) Column and row headings should be brief but self-explanatory.
(iv) Units of measurement should be shown clearly.
(v) Approximations and omissions can be explained in footnotes.
(vi) A vertical arrangement of figures is generally preferable to a horizontal arrangement because columns of figures are more usual than rows.
(vii) Double lines, or thick lines, can be used to break up a large table and make it easier to read.
(viii) Two or three simple tables are often better than one very large table.
(ix) Sets of data which are to be compared should be close together.
(x) Derived statistics, such as percentages and averages, should be beside the figures to which they relate.

Tables re-present data in order to emphasise certain features and to omit irrelevant detail, therefore it is important to be selective about the data that is put into a table and the number of tables produced.

When there are only a few variables involved tabulation can be carried out by hand quickly and easily. However, when there are a number of variables and combinations required with cross-tabulations, then the information can be fed into a computer. A computer will produce tables very quickly in large volumes if required and in a short time tables can be produced that can involve weeks of work to analyse and interpret. Therefore it is important to be clear about the aims and objectives of presenting the data before vast numbers of useless tables are produced.

For example: if information is required about total unemployment among school-leavers, it would be quite possible to produce tables showing a large number of combinations of age against sex and type of school and father's occupation. However, if the information required is simply the total figure to compare with other years, then all the other possible tables would be unnecessary.

7.3 CLASSIFICATION

Tabulation involves classification, and before data can be tabulated, interpreted and presented it must be classified. Classification is the process of relating the separate items within the mass of data collected and the definition of various categories in the table.

Every piece of data has characteristics, some of which are measurable attributes or variables (such as weight) and some of which are non-measurable attributes (such as beauty). Measurable variables can be divided into:

(a) Discrete variables
These are measured in single units (such as people, houses, cars). However, statements are made about discrete variables which appear to contradict this. For instance, 'the average family has 2.3 children' and 'on average people live in 1.2 rooms'. In fact it is difficult to think of a fraction of a child or a room, but when figures are averaged these results are possible.

(b) Continuous variables
Continuous variables are in units of measurement which can be broken down into definite gradations. Examples include temperature in decimals of a degree, height or length in decimals of a centimetre or fractions of an inch. However, in practice continuous variables are converted to a discrete form by expressing values to the nearest appropriate unit of measurement. It is difficult and unnecessary to distinguish small differences.

(c) Class intervals
In producing a table of a frequency distribution class intervals have to be shown clearly and unambiguously. There are many different methods of doing this.

Where a discrete variable is involved the following method can be used:

    (i)   Number of people
            100-199
            200-299
            300-399
            400-499

There cannot be any confusion between these class intervals. It is clear that the 199th person will be in the first class, and the 200th person will be in the second class. Other forms used include:

(ii)	No. of people	(iii)	No. of people
	100-200		100-
	200-300		200-
	300-400		300-
	400-500		400-

These class intervals are perhaps less clear. In (ii) it is not clear into which class 200 should fall, or 300 or 400. In each case there are two classes which could include these figures. In (iii) it has to be inferred where the classes end, particularly the last class.

With continuous variables the level of approximation or rounding needs to be shown clearly. For example, the height of people can be shown to very precise levels of measurement such as a millimetre or a fraction of an inch. In practice these heights are likely to be the nearest centimetre or to the nearest inch:

    (i)   Height (to the nearest centimetre)
            1.40 metres-1.59 metres
            1.60 metres-1.79 metres
            1.80 metres-1.99 metres

Within these class intervals, someone who is 1 metre 59 centimetres tall will be in the first class. Somebody else who is 1 metre 59.5 centimetres will be in the second class. The 'rules' of approximation and rounding are followed.

These class intervals can be shown in the form:

(ii)	Height	(iii)	Height
	1.40 m-1.60 m		1.40 m but less than 1.60 m
	1.60 m-1.80 m		1.60 m but less than 1.80 m
	1.80 m-2.00 m		1.80 m but less than 2.00 m

These forms show continuous variables, but not the extent of the approximations used. 1.40 m but less than 1.60 m could mean up to 1 metre 59.99 centimetres or up to 1 metre 59.49 centimetres. However, these forms can be used if the level of approximation is given.

(d) Open-ended class intervals

This is another problem of class-interval construction. Open-ended classes normally occur at the beginning or end of a frequency distribution.

For example:

>   Age of employees
>   (in years)
>   under 20
>   20 but under 40
>   40 but under 60
>   60 and over

In deciding the extent of the open-ended class interval (which is often necessary to carry out calculations), there are a number of points to be considered:

(i) The extent of adjacent class intervals. In the example, because the third class interval is 40 but under 60, it could be argued that it would be consistent to make the last class interval 60 but under 80.
(ii) There may be practical factors involved in classification, such as if the school-leaving age is 16 then the probability is that the first class will be 16 but under 20. Also in the example, if the firm involved strictly enforced retirement at 65, the upper class would be 60 but under 65.

Decisions on the extent of open-ended class invervals are a matter of judgement, so it is useful to state (in a footnote or in the report) the reasoning behind the decision which has been taken.

7.4 FREQUENCY DISTRIBUTIONS

These show the frequency with which a particular variable occurs.

For example: a traffic survey may show the following flow of vehicles passing a particular point during an hour:

Vehicles	Frequency
Cars	45
Lorries	22
Motorcycles	6
Buses	3

In practice the observer may have had a list of these types of vehicles and put a line or mark against the appropriate category when a vehicle passed:

Vehicles		Frequency
Cars	|||| |||| |||| |||| |||| |||| |||| |||| ||||	45
Lorries	|||| |||| |||| |||| ||	22
Motor cycles	||||		6
Buses	|||	3

Another example: a survey of the number of people per household in a particular street of 20 houses may produce the following results:

Households	No. of people	Households	No. of people
A	4	K	3
B	3	L	2
C	2	M	3
D	6	N	2
E	3	O	2
F	1	P	3
G	4	Q	5
H	3	R	4
I	2	S	5
J	2	T	4

These results can be put into a table in the form of a frequency distribution:

No. of people per household	No. of households
1	2
2	5
3	6
4	4
5	2
6	1
	20

This is a grouped frequency distribution with the frequencies grouped according to the number of people per household.

In frequency distributions a decision has to be made about the grouping or classification used. This is not a problem in the above example, but it is more difficult with very large amounts of data.

For example: if a table is constructed from the wage list of a firm employing 1000 people it may be necessary to produce a summary in the form of a frequency distribution (Table 7.1):

Table 7.1 frequency distribution: Wage list of a firm employing 1000 people

Classes (weekly wages, to the nearest £)	Frequency (no. of employees earning wages within these classes)
40 to 59	15
60 to 79	40
80 to 99	140
100 to 119	620
120 to 139	150
140 to 159	35
	1000

This is a grouped frequency distribution with the variable values grouped into intervals to provide a summary which clarifies the distribution of wages within the firm. For instance, it can be seen from table that 91% of the employees earn between £80 and £139 a week, with only a relatively few people earning less or more than these limits. The 15 employees earning less than £60 a week are poorly paid compared with the rest, while the 35 employees earning £140 or more a week are well paid.

From a table like this one it is possible to see a structure in the figures which might be more difficult to identify by looking at a list of a thousand wages.

7.5 REPORTS

These form an early stage of presentation, produced alongside or immediately after tables have been constructed.

Reports can be used:

(i) To explain the background to the collection of data. The way the information has been collected can be detailed, whether it has been as a result of a primary survey or come from secondary data.
(ii) To explain reasons for producing some tables and not others.
(iii) To interpret the information contained in tables and other forms of presentation.
(iv) To emphasise points of importance.

Textual reports are often the simplest method of presenting data and the easiest to understand for people not used to assimilating facts from tables. However, reports can include too many figures to be easily assimilated, so that a combination of tables with a report may provide the clearest form of presentation.

The main features of a good report are:

(i) accuracy,
(ii) brevity,
(iii) clarity.

7.6 HISTOGRAMS

Once tables have been constructed and a report written, the next stage in presentation is to produce diagrammatic illustrations of the data.

The histogram is a method of representing a frequency distribution diagramatically.

The word 'histogram' is derived from the Greek *histos*, or mast, so that it is a mast diagram or bar diagram. A histogram consists of a series of blocks or bars each with a width proportional to the class interval concerned and an area proportional to the frequency.

Fig 7.2 *line chart*

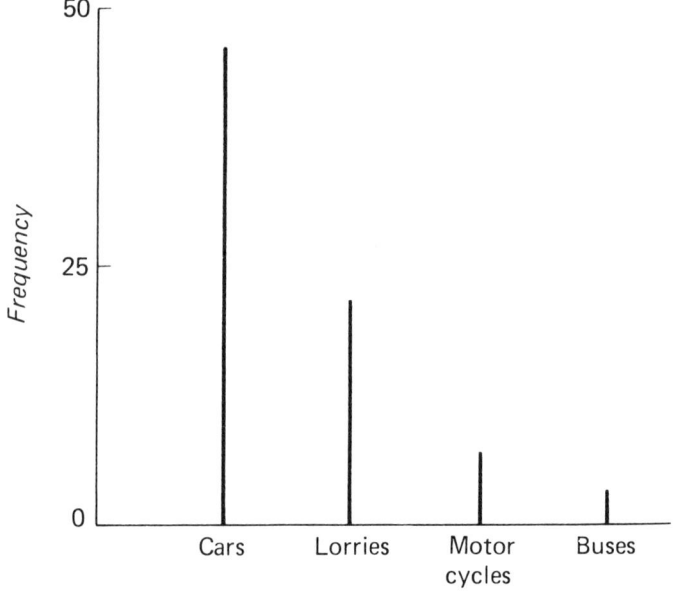

(a) **Line charts**: these are the simplest diagram used to represent a frequency distribution, where the length of each line is proportional to the frequency.

For example: in the traffic survey in Section 7.4 the observations showed:

Vehicles	Frequency
Cars	45
Lorries	22
Motor cycles	6
Buses	3

A line chart can be drawn to represent this data. The length (or height) of each line represents the frequency (Figure 7.2):

(b) **Histograms**: the histogram is an extension of the line chart.

For example: in a survey of the output of machine operatives in a factory, the results shown in Table 7.2 were obtained:

Table 7.2   **output of machine operatives**

Output (units per operative)	Number of operatives
200-209	5
210-219	14
220-229	17
230-239	29
240-249	42
250-259	21
260-269	10
270-279	2

In a histogram this frequency distribution would appear as shown in Figure 7.3.

Whereas in the line chart the length of the line was proportional to the frequency, it is the areas of the blocks which must be proportional in the histogram.

This can be seen by amalgamating the last three groups in the frequency distribution. This will make a class interval between 250 units and 279 units with a frequency of 33 operatives. This is shown by drawing a block which is three times the width of the others and a third the height (Figure 7.4).

Fig 7.3 *histogram*

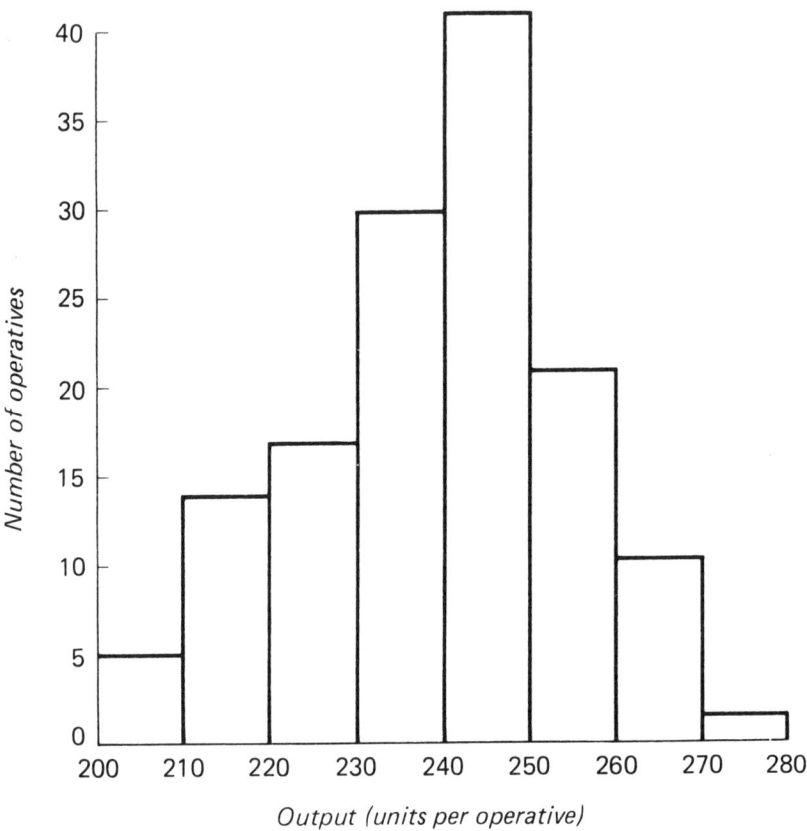

Output (units per operative)

The area of a histogram bar is found by multiplying the width of the bar (the class interval) by the height (the frequency). Therefore, in Figure 7.4:

the area of the last block is:
30 x 11 = 330
the area of the last three blocks in Figure 7.3 are:
10 x 21 = 210
10 x 10 = 100
10 x 2 = 20
           ———
           330
           ———

Therefore, if the areas of the blocks are to maintain their proportional relationship with the frequencies, the block height and width must be

Fig 7.4 *histogram with uneven class intervals*

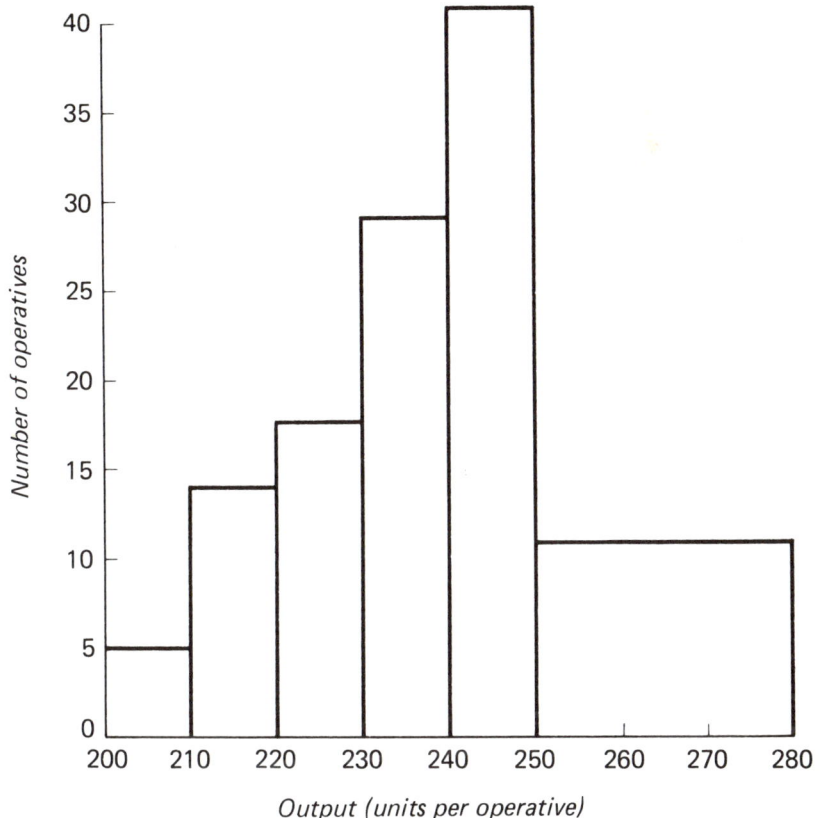

Output (units per operative)

changed proportionately. In figure 7.4, therefore, the last block height is reduced to a third of the frequency and the numerical width of the block is trebled. It follows that whenever a grouped frequency distribution includes uneven class intervals, the area of the blocks must be kept in proportion.

(c) **the construction of a histogram** involves the following points:

(i) The horizontal axis is a continuous scale including all the units of the grouped class intervals.
(ii) For each class in the distribution a block (or vertical rectangle) is drawn extending from the lower class limit to the upper limit.
(iii) The area of this block will be proportional to the frequency of the frequency of the class.
(iv) If the class intervals are even throughout a frequency table, then the

height of each block is proportional to the frequency.
(v) There are never gaps between histogram blocks because the class limits are the true limits in the case of continuous data and the mathematical limits in the case of discrete data.

7.7 FREQUENCY POLYGONS

This is a diagram which is drawn by joining up the mid-points of the tops of histogram blocks. This is usually drawn with straight lines and the resulting diagram taken to include the horizontal axis is a 'many-sided figure' or polygon.

For example: see Table 7.3 and Figure 7.5.

Table 7.3   **overtime pay**

Overtime pay (to the nearest £)	Number of employees
0-4	4
5-9	10
10-14	5
15-19	4
20-24	2
	25

In the case of the first and last class intervals, the line can be extended beyond the original range of the variable. This is because the area under the polygon should be the same as that in the histogram. Only if each triangle cut off the histogram is compensated for will this requirement be met. This can be seen in Figure 7.5 by the lettered triangles.

In theory a frequency polygon is always drawn by constructing the histogram first; in practice, each point can be located by reference to the frequency and the mid-point of the class interval. Therefore, using the data from Table 7.3, the frequency polygon in Figure 7.6 can be constructed.

Histograms and frequency polygons enable the properties of distributions to be examined and various forms of distribution can be compared in a general way (see Chapter 10 for comparisons). However, neither the histogram nor the frequency polygon gives a very accurate picture of a frequency distribution. The histogram implies that the frequencies are the same throughout the class interval, when they may not be; the frequency polygon implies that sharp, angular differences occur between the mid-points of the class intervals when this may not be true.

Fig 7.5 *frequency polygon*

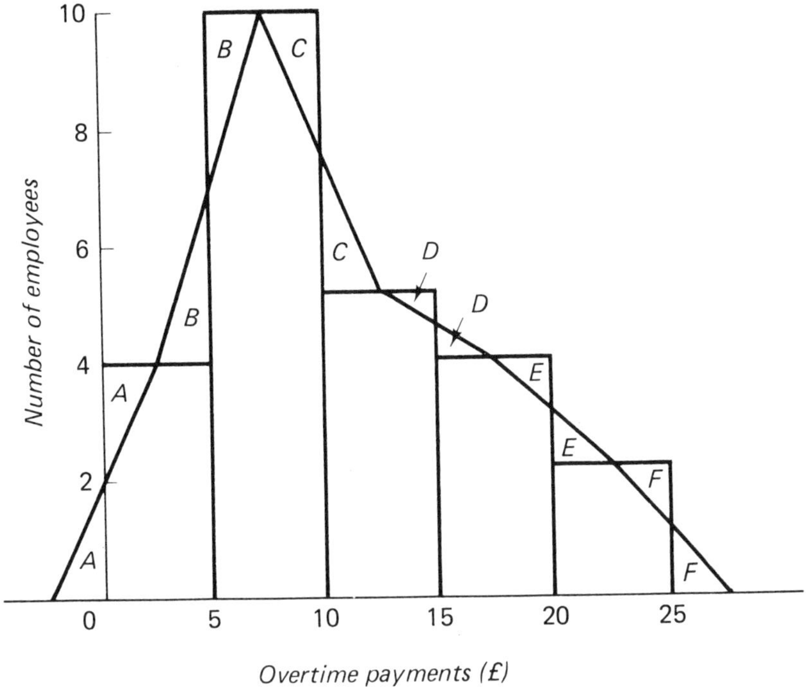

Overtime payments (£)

Fig 7.6 *overtime pay*

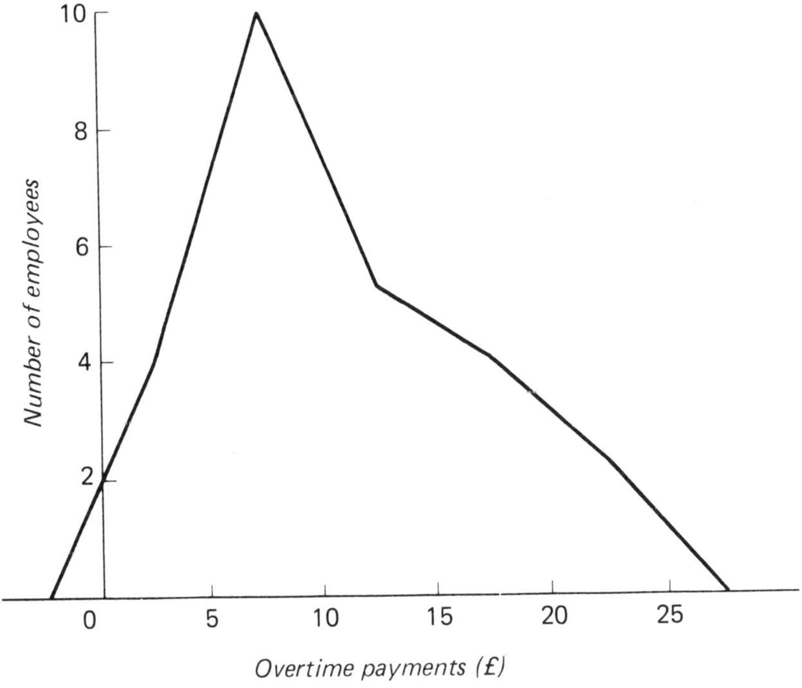

Overtime payments (£)

It is likely that a more accurate idea of the distribution would be obtained if there were smaller class intervals and more observations. In other words the closer the class intervals are to the original set of figures the greater the accuracy obtained. This may be at a cost of a loss of summarisation and clarity.

7.8 FREQUENCY CURVES

As the number of class intervals and frequencies are increased so the polygon and histogram move more closely towards a curve (Figure 7.7):

Fig 7.7 *a frequency curve*

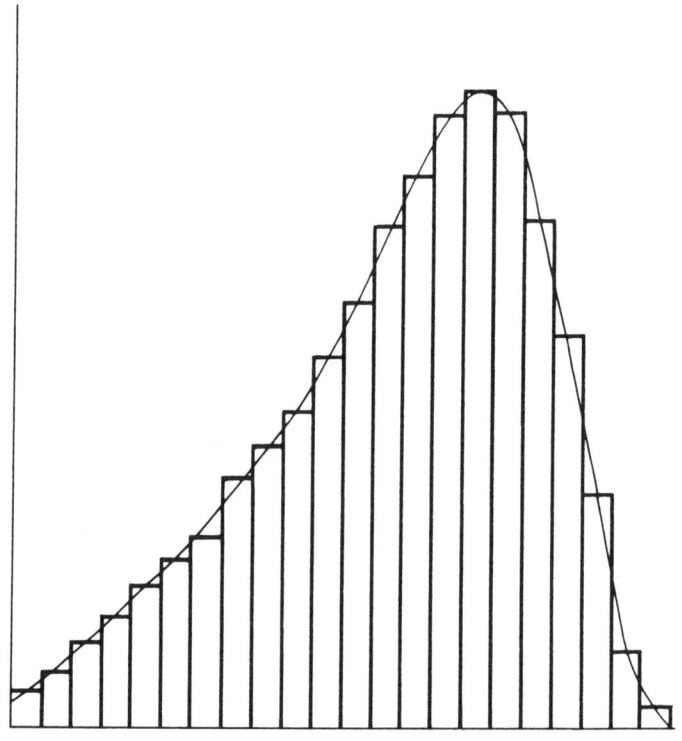

A frequency curve is formed by:

(i) smoothing out a histogram or frequency polygon, or
(ii) using smaller class intervals and more observations to smooth out the line of the curve.

The importance of the frequency curve, like the histogram and polygon, is that it can provide a clear picture of the 'shape' of a distribution. With a small distribution the shape of a distribution may be seen easily by looking at the figures, but with large amounts of data it may be more difficult. The frequency curve can provide a summary at a glance and make an immediate impact.

More than one frequency curve can be plotted on the same axis for comparison.

For example: see Figure 7.8

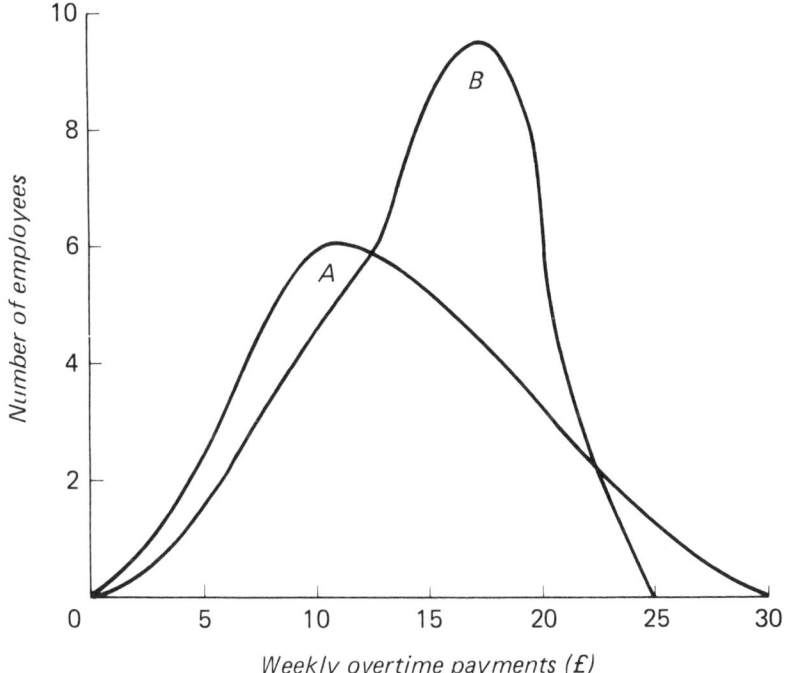

Fig 7.8 *weekly overtime payments by two companies*

From the two curves it can be seen that:

(i) In general company A made lower overtime payments than company B.
(ii) The highest level of overtime is paid by company A, as well as the lowest.
(iii) Company B has a greater number of employees earning overtime than

company A (the area under curve B appears to be larger than the area under curve A).
(iv) Most of company B's overtime payment is made at the top end of the overtime scales, while company A's overtime payments are at the lower end of the scale.
(v) There is some overlap of overtime payment.

ASSIGNMENTS

1 Draw a histogram from the data in Table 7.1. Briefly, comment on the distribution.

2 Write a short report on the problems of classification.

3 Comment on the differences in the two distributions shown in Figure 7.9.

Fig 7.9

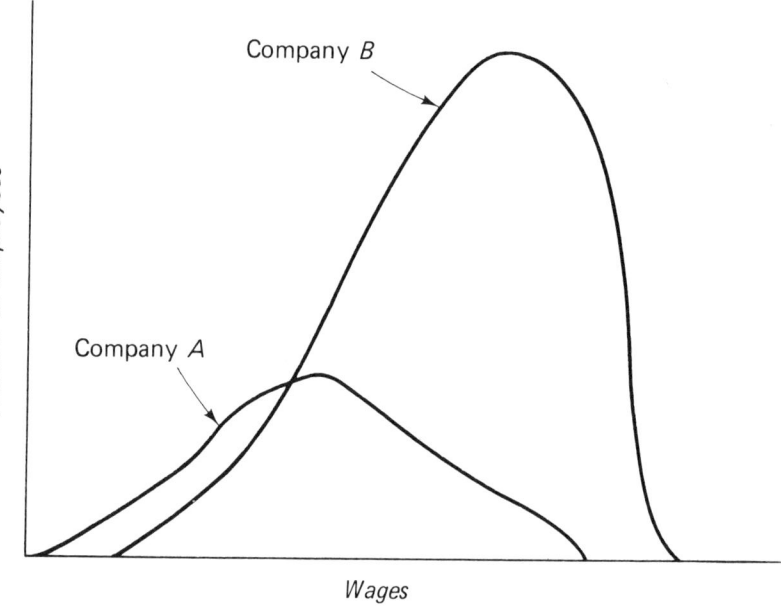

4 Discuss, with examples, the aims of statistical presentation.
5 (i) Draw up a table to present the information that is likely to be collected from Assignments 4.2 and/or 4.3.
   (ii) Draw up a table to show the main tasks carried out during a person's working day and the time spent on each task.

CHAPTER 8

HOW TO PRESENT FIGURES (2)

Tables, reports and frequency distributions are the basic methods of presenting raw data. The next stage is to present data in a pictorial form which will make an immediate impact, illustrate the information and bring out the salient points. Pictorial presentation falls into two main categories:
(i) charts,
(ii) graphs.

8.1 BAR CHARTS

Bar charts are amongst the most popular forms of pictorial presentation. There is a range of commonly used bar charts:

(a) **The simple bar chart**: see Figure 8.1.

This chart makes comparison between the years easy. The height of each bar shows the total output for each year. This chart should not be confused with the histogram. It is more like a line chart than a histogram (see Section 7.6) because it is the height of each bar which is important. The height (or length) represents the data, the width and area of the bar are not important because they are not drawn in proportion to any data as they are in the histogram. For this reason all the bars on any particular bar chart are the same width.

Simple bar charts (like Figure 8.1) can be used to illustrate only simple pieces of information but can illustrate information clearly and provide an immediate visual impact. Figure 8.1 shows that the total output of Company A has increased over the three years but has increased more slowly between 1982 and 1983 than between 1981 and 1982.

The bar chart can be drawn with horizontal bars (Figure 8.2), if this improves the visual impact of the chart.

Fig 8.1 *the simple bar chart*

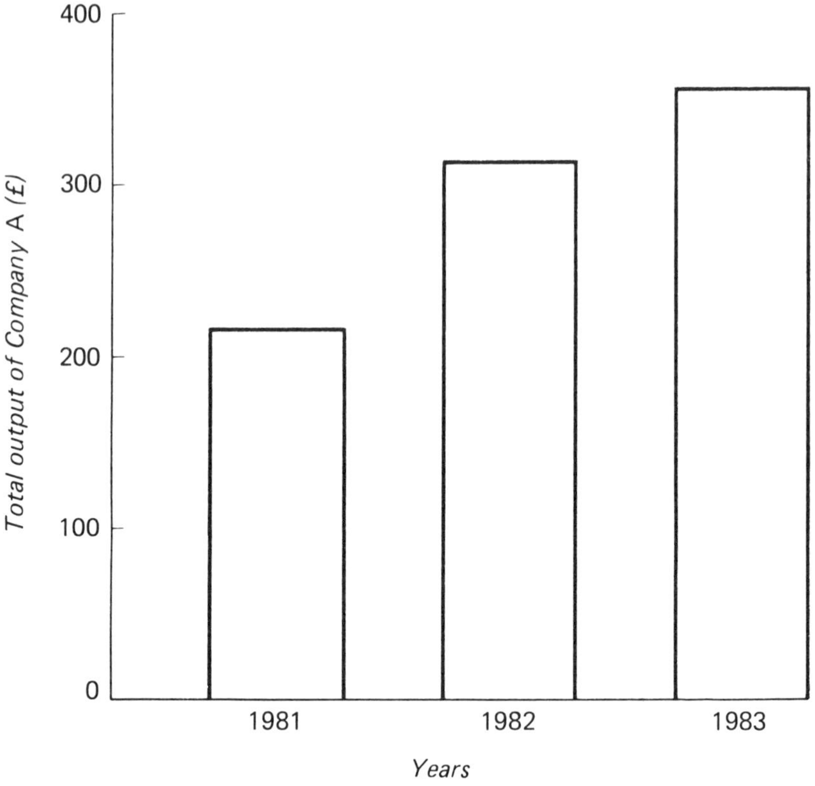

Fig 8.2 *horizontal bar chart: changes in output over two years*

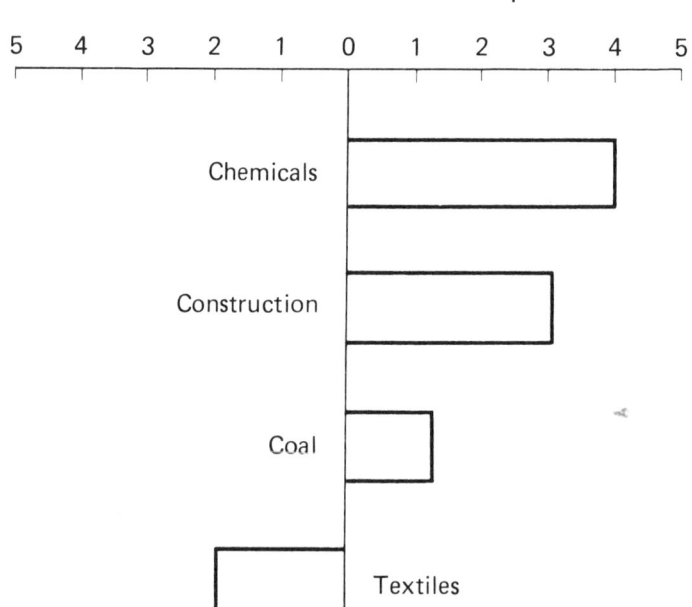

(b) **The compound bar chart**: this form of bar chart (see Figure 8.3) is useful for comparing a number of items within say a year, as well as comparing the items between the years.

Fig 8.3 *the compound bar chart: output of an electrical goods company*

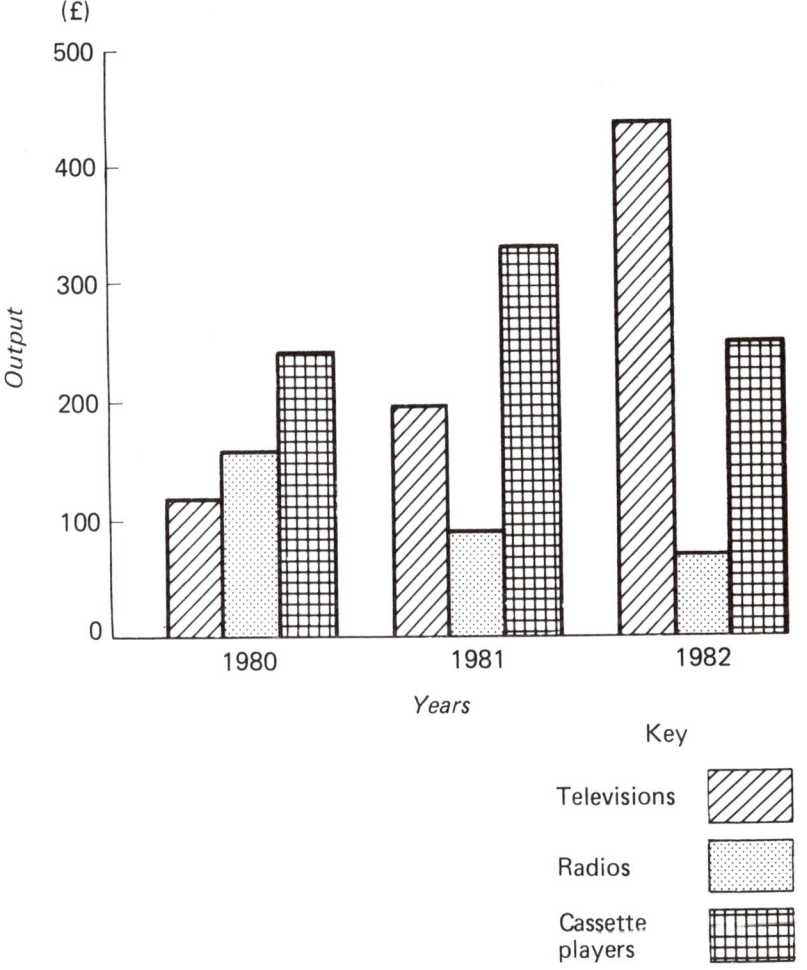

(c) **Component bar charts**: these are useful to show the division of the whole of an item into its constituent parts (see Figure 8.4).

Figure 8.4 shows the output of an electrical goods company over three years. The chart shows that total output in each year has risen. The output of television sets has increased but output of radio sets has declined while the output of cassette players has fluctuated in each year.

This type of chart enables a comparison of the total output to be made easily, while the compound bar chart emphasises the comparisons between items.

Fig 8.4 *component bar chart: output of an electrical goods company*

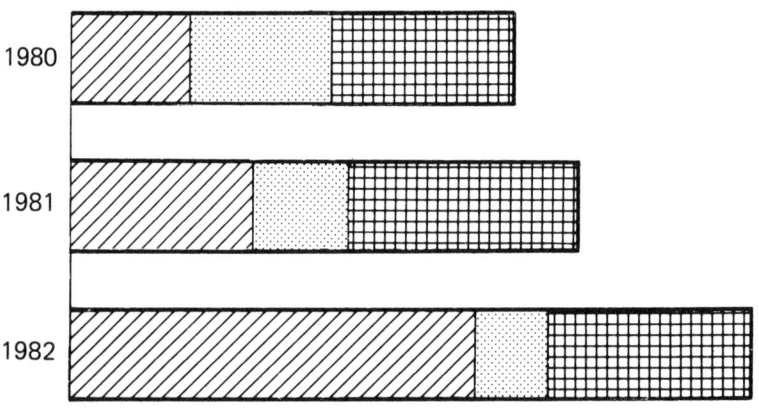

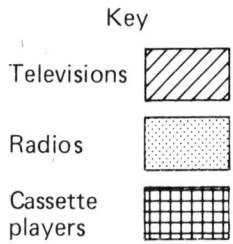

(d) **Percentage component bar chart**: in this chart (Figure 8.5) the bars represent 100% and therefore remain the same length. The components change to represent the percentage they make up of the total.

Fig 8.5 *percentage component bar chart: output of an electrical goods company*

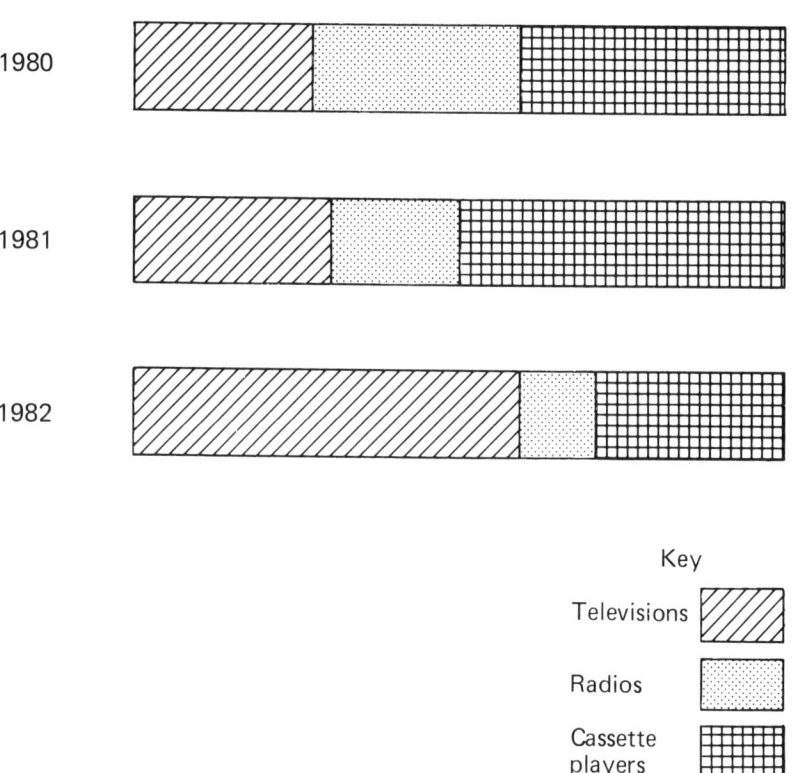

This chart shows the changes in the components of total output, but does not show the fact that total output has risen over the three years. The chart shows that televisions have become a larger proportion of total output, while radios and cassette players have become a smaller proportion.

8.2 PIE CHARTS

A pie chart is a circle divided into sectors to represent each item or variable (pie or pi from the Greek π, πr^2 the area of a circle, or pie as in apple). Each sector of the circle should have an area equal to the quantity of the variable.

For example: if the quantity of a variable is 15 units out of a total of 45, then the variable must occupy $\frac{15}{45}$ ths of the area of the circle. In total the area of a circle includes 360°, so therefore this variable will occupy:

$$\frac{15}{45} \times 360 = 120°$$

Therefore, if the finances of a company are:

	£ m.
Costs	20
Profit	10
Taxes	15

The circle is divided in proportion:

$$\frac{10}{45} \times 360 = 80° \qquad \frac{20}{45} \times 360 = 160°$$

The three sectors or segments of the circle will be 120°, 80°, and 160° (Figure 8.6).

Fig 8.6 *pie chart: company finances*

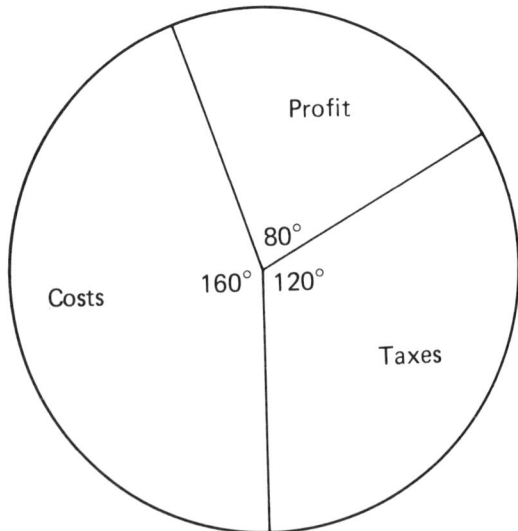

Often pie charts are presented with the areas shaded or coloured and a key provided (Figure 8.7):

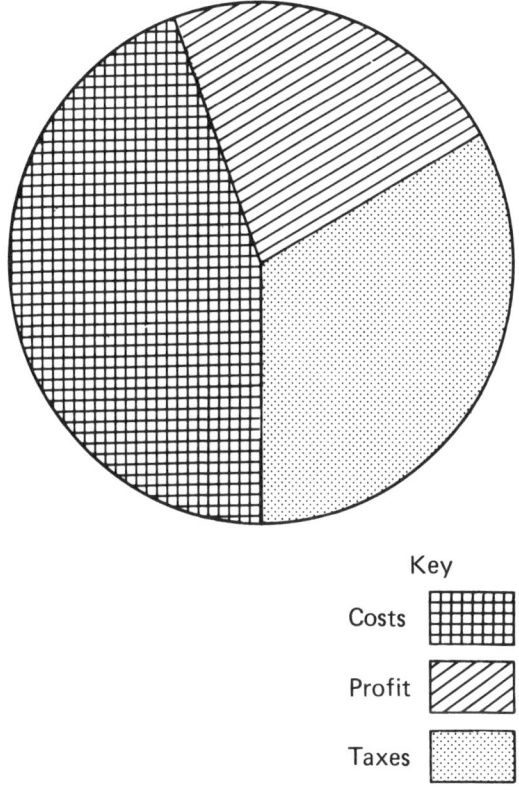

Fig 8.7 *shaded pie chart : company finances*

Pie charts are useful where there are a few items which make up proportions of a whole and where the proportions are more important than the numerical values. For instance, shareholders may be more interested in the proportion of profit than the actual values; they want to know the size of their share of the cake.

Pie charts are useful when:

(i) There are few variables to be included.
(ii) These variables make up proportions of a whole figure.
(iii) The numerical values are less important than the proportions of the area of the whole circle.
(iv) The aim is to provide a strong visual impact.

Pie charts raise problems because:

(i) They can involve long calculations (see comparative pie charts).
(ii) They do not provide information on absolute values unless figures are inserted in each segment.
(iii) The segments cannot be scaled against a single axis as the values can on a bar chart.
(iv) To compare two totals the areas of the pie charts should be in proportion; the fact that the area of one circle is larger than the area of the other circle may not be very clear visually.

8.3 COMPARATIVE PIE CHARTS

These are used to compare two sets of data, usually two sets of the same items over time. The areas of the circles must be in proportion to the totals of the data.

For example: finances of a company over two years:

	1980/81 (£m.)	1981/82 (£m.)
Costs	20	20
Profit	2	10
Taxes	8	15
	30	45

A pie chart is constructed for 1980/81 with a radius of say 2 cm. The area of a circle is πr^2. Therefore $r^2 = 4$ cm: £30 m is represented by a circle of area π 4 square centimetres (cm^2); £10 m would be represented by a circle of area π 4 ÷ 3 = π 1.33 cm^2; £45 m would be represented by a circle of area π 1.33 × 4.5 = π 6 cm^2.

If the radius of the second pie chart representing 1981/82 is r, the area in cm^2 will be πr^2. The area has to be π 6 cm^2. Therefore

$$\pi r^2 = \pi 6$$
$$r^2 = 6$$
$$r = 2.4 \text{ cm}$$

There is no need to substitute the numerical value for π.

Therefore the first pie chart will have a radius of 2 cm and the second pie chart a radius of 2.4 cm with the appropriate segments for the three variables (Figure 8.8):

While the sectors or segments can be compared relatively easily, the areas of the two circles are too close to each other to make comparison easy.

Fig 8.8 *comparative pie charts: company finances over two years*

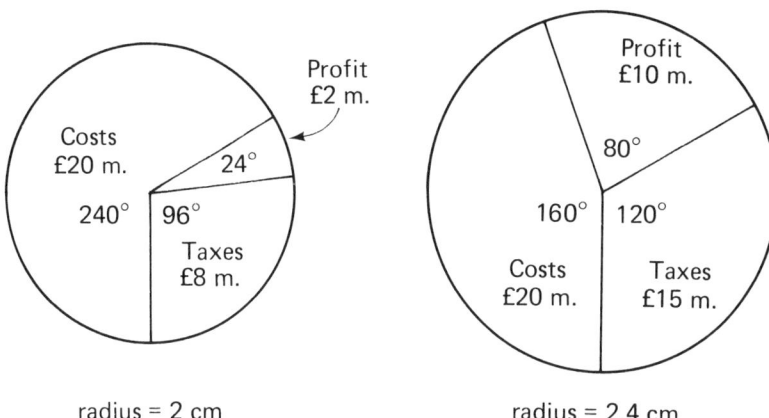

8.4 PICTOGRAMS

Pictograms are *picto*rial dia*grams*. They are pictures to represent data. The objective in using them is to provide an immediate visual impact and therefore they should be kept simple. Pictograms cannot give detailed information, but can show trends, comparisons and totals.

For example: Figure 8.9.

This pictogram does show that the number of cars produced by this company is falling and at the same time the number of workers is falling at the same speed between 1970 and 1975, but faster between 1975 and 1980. In 1970 the ratio of workers to cars produced was 3 : 1, in 1975 it was 3 : 1 and in 1980 it was 2 : 1.

Pictograms can:

(i) Be an attractive method of representing data.
(ii) Make it easy to appreciate fluctuations in variables.
(iii) Make an immediate visual impact.

Pictograms may:

(i) Not make it easy to estimate the precise numbers involved.
(ii) Be misleading unless the symbols used are always the same size. It is easier to compare variables by increasing the number of symbols rather than the size or area of the symbol.
(iii) Be used only for showing relatively simple information.

Fig 8.9 *pictogram: number employed compared with the output of cars*

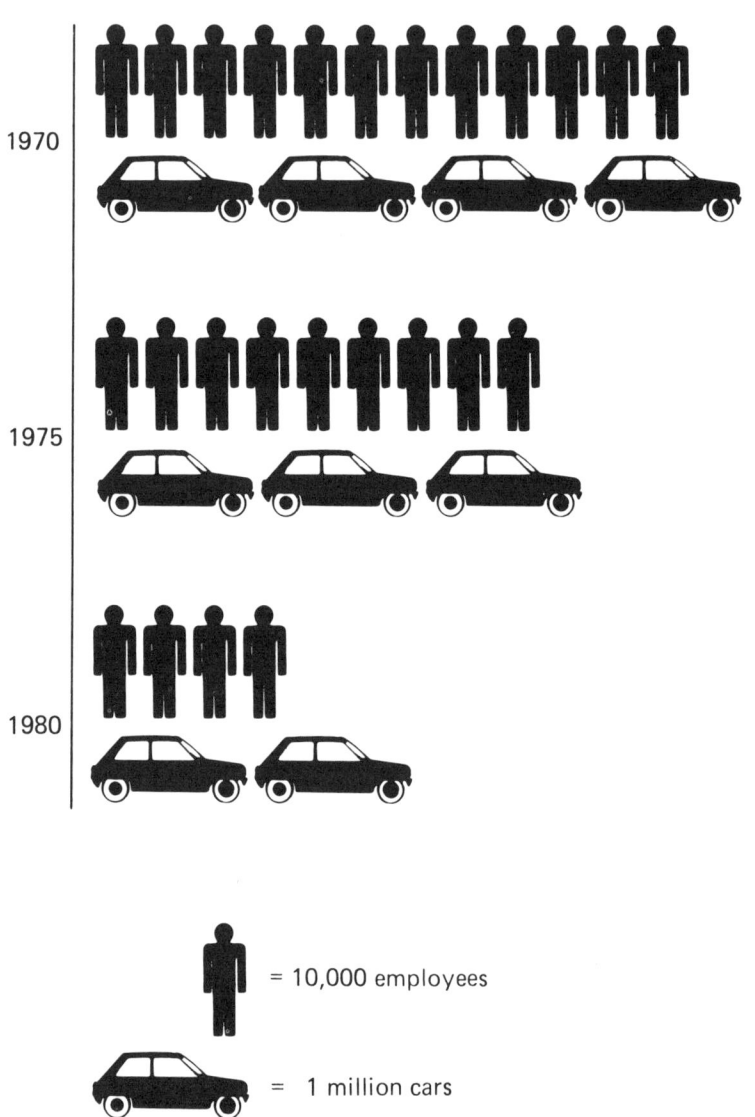

8.5 COMPARATIVE PICTOGRAMS

These are pictograms which can be used to compare relationships by using symbols and pictures of different sizes. As in the pie chart, the area of the symbol should be in proportion to the quantity of the items being compared.

For example: Figure 8.10.

Fig 8.10 *comparative pictogram*

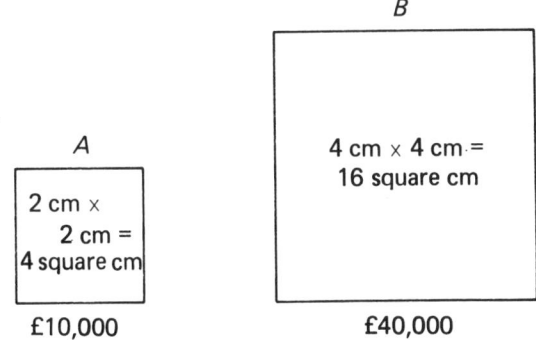

Square B is four times the area of square A, to represent a four times larger amount of money.

8.6 CARTOGRAMS OR MAP CHARTS

These are maps onto which are superimposed graphs, symbols, pictograms, flags and so on to represent various factors. For instance, maps are used by salesmen, with flags or other symbols to show outlets for their goods. Also, cartograms are used to show the main industries of a country and the main imports and exports.

Any of the other forms of presentation discussed in this chapter can be superimposed on to a map where this is appropriate. Cartograms can provide an attractive and easily understood visual form of representation where geographical location is important.

For example: Figure 8.11.

Fig 8.11 *cartogram: company regional sales areas in England and Wales*

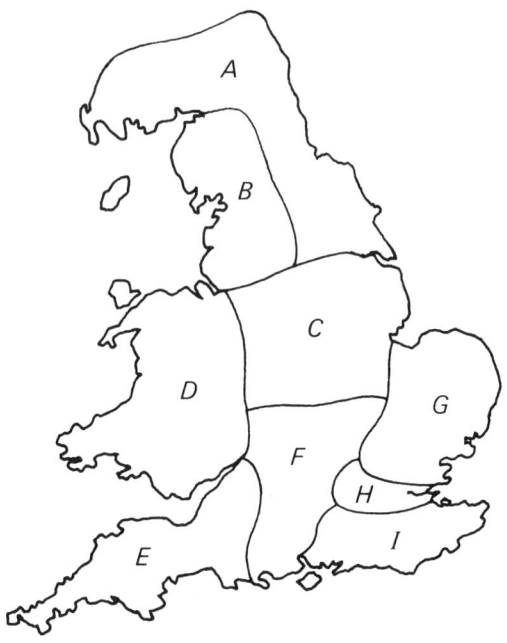

8.7 STRATA CHARTS

These may also be called layer charts, or band curve graphs, because each band or strata is placed successively on top of the previous one. Totals are cumulative, so that each element of the whole is plotted one above another.

For example: Figure 8.12.

In Figure 8.12 the top line shows the total production costs. It is relatively easy to see the amount made up by each layer. Overheads are a larger proportion of total costs than labour or raw materials by 1981. It is more difficult to compare layers or strata which are close to each other in height, or to read off the amounts for each item.

These charts can provide immediate evidence of the relative importance of the various constituents of the total expenditure.

8.8 GRAPHS

In theory a graph consists of a grid with four quadrants with zero (the origin) at the centre. Curves are drawn on the grid to illustrate the relationship between two variables (Figure 8.13).

Fig 8.12 *strata chart: production costs of company* A

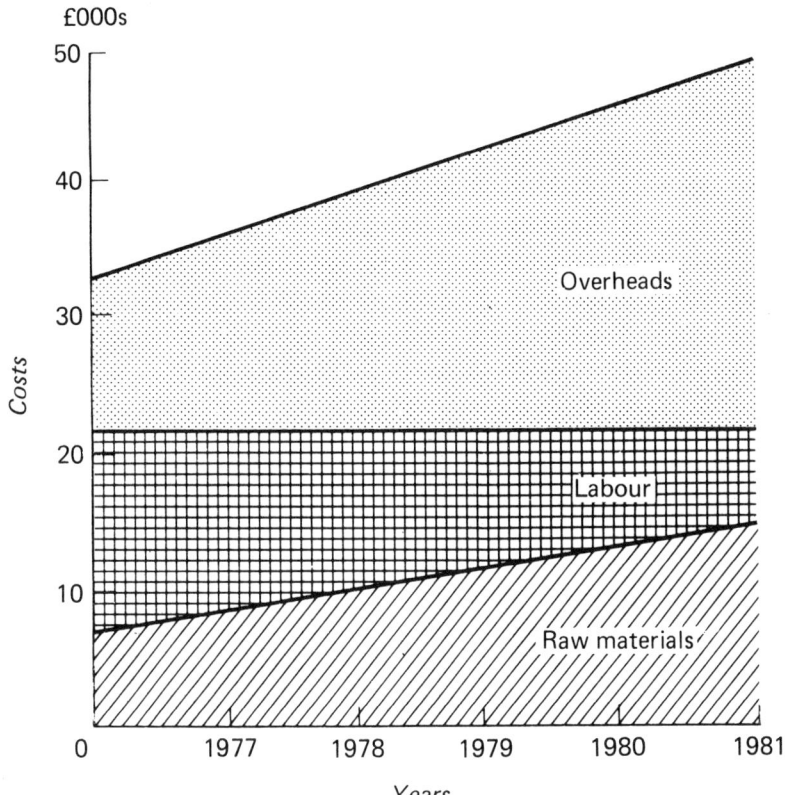

Usually the negative axes are not drawn, because although variables can be negative as well as positive, they are usually positive. A graph is a form of pictorial presentation and the word can be associated with the word 'graphic' or a 'vivid representation'.

The vertical axis or y axis (the positive vertical axis), is scaled in units of the dependent variable. The horizontal axis or x axis (the positive horizontal axis), is scaled in units of the independent variable. *The independent variable* is the variable which is not affected by changes in the other variable. *The dependent variable* is the variable which is affected by changes in the other variable.

For example: changes in advertising expenditure may affect sales, but the level of sales will not directly affect advertising expenditure. Therefore advertising is the independent variable and sales the dependent variable (Figure 8.14).

Fig 8.13 *the four quadrants of a graph*

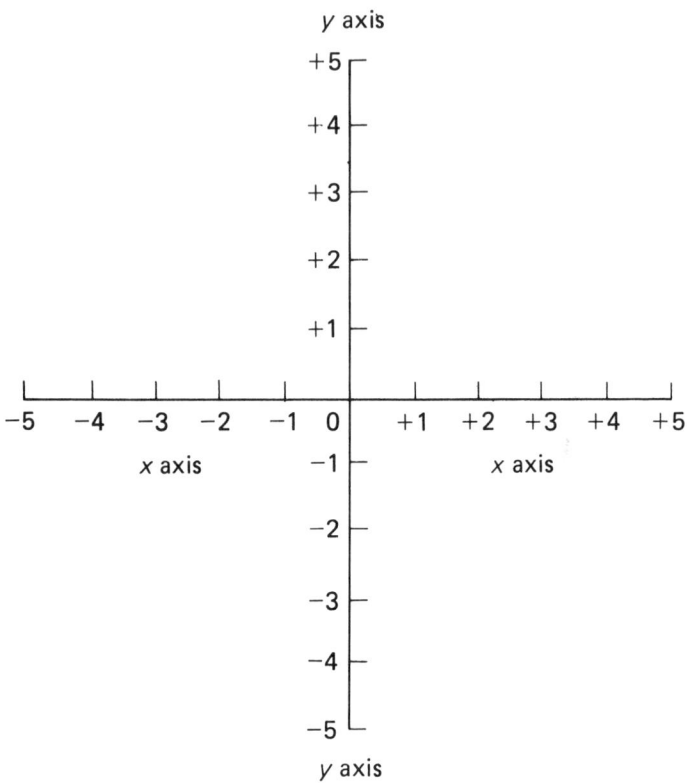

In practice it is not always easy to decide which variable is dependent and which independent. For instance, changes in the costs of a firm may affect output and output may affect costs. From different viewpoints they may be dependent on each other.

Normally the independent variable is chosen before the information is collected and therefore it is stated in units or class intervals or periods of time. The dependent variable is the number or frequency in each class interval or period of time.

The curve of a graph is usually drawn as a freehand curve, but may be drawn as a series of straight lines. Strictly, if the dependent variable is discrete then the points should be joined by straight lines. If the dependent variable is continuous then the points should be joined by a smooth curve.

Fig 8.14 *a graph*

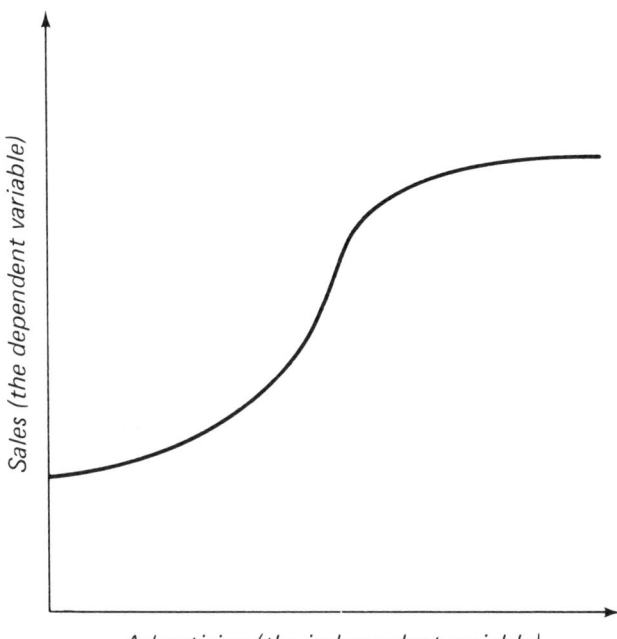

Advertising (the independent variable)

In fact a smooth curve is often used with discrete as well as continuous variables. However, it is only with continuous variables that values between points can be read off (interpolated), because with a discrete variable there is no information between one point and the next.

Also, with continuous variables it is possible to extrapolate by finding a value outside the given range, by extending the graph on the assumed trend (see Chapter 14).

The position of any point on the curve of a graph is decided by reference to the axes. The points where the variables intersect are called 'bearings' or 'co-ordinates'. A set of points is built up and these are joined to form the curve. Like reading a map, on a graph a point is fixed by reading out from the origin (where the x and y axis intersect at 0) along the horizontal axis and drawing a vertical line up from this point and then reading up from the origin on the vertical axis and drawing a horizontal line along from this point (as in map reading this can be described as: 'along the corridor and up the stairs').

For example: Table 8.1 and Figure 8.15.

Table 8.1   **monthly sales**

Months	Sales of Company *A* (£000s)
January	100
February	250
March	400
April	480
May	420
June	340

Fig 8.15 *sales graph: sales of company* A

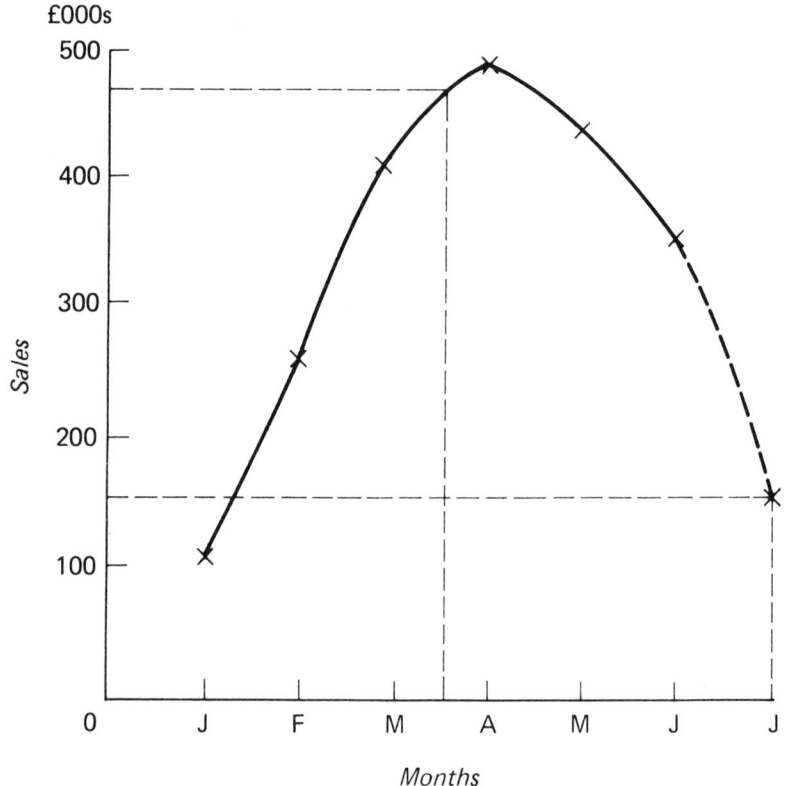

From Figure 8.15 it is possible to interpolate that the level of sales in the middle of March was at a level of about £450,000. If it is assumed that the downward direction of the curve continues then it is possible to extrapolate that sales will be down to a level of £150,000 by July. However, this is a dangerous assumption because the trend could have changed due to a range of factors and the period of the graph does not provide sufficient evidence of the trend (see Chapter 14).

The vertical line should always start at zero, otherwise a false impression may be created. If it is not practical to have the whole scale running from zero then the scale can just cover the relevant figures providing that zero is shown at the bottom of the scale and a definite break in the scale is shown. This break can be shown in these ways:

(i) The vertical line can be broken or zigzagged to show that some of the empty space has been compressed (Figure 8.16).
(ii) The vertical scale can be broken by two jagged lines running across the diagram to indicate that a portion of the space has been omitted (Figure 8.17).

When the method of compiling or calculating the figure under review has been changed, this should be shown by a break in the axis and in the graph (Figure 8.18). This happens with index numbers when a new series is developed with a new base date and weighting system (see Chapter 12).

Fig 8.16 *broken scale (i)*

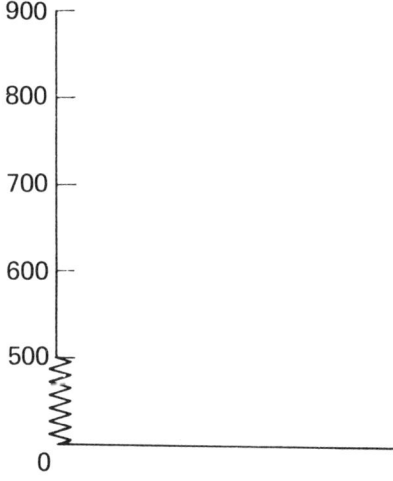

Fig 8.17 *broken scale (ii)*

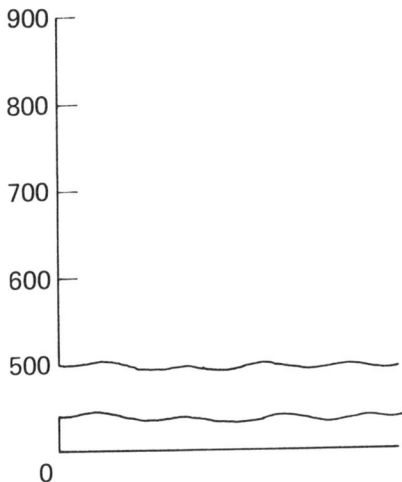

Fig 8.18 *broken curve*

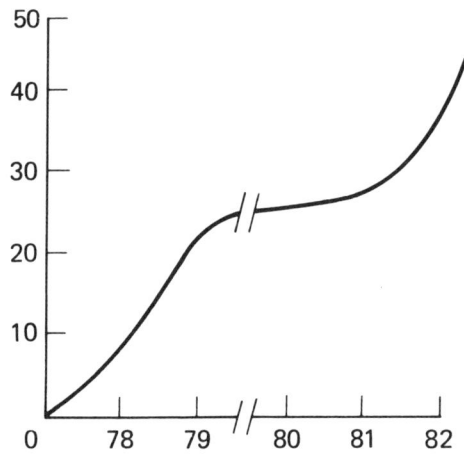

8.9 SEMI-LOGARITHMIC GRAPHS

The semi-logarithmic graph (or semi-log or ratio scale graph) is used to show the rate of change in data, rather than changes in actual amounts (which are shown on natural scale graphs).

Usually only one axis (the y or vertical axis) is measured in a logarithmic or ratio scale. Therefore the graph is called a *semi*-log graph.

The important factor on a semi-log graph is the degree of slope of the curve. The curve of the usual graph measures the magnitude at any point, while the log graph shows at any point the percentage change from the last point.

For example: Table 8.2, Figure 8.19 and Figure 8.20.

Table 8.2 **semi-log table**

Years	Profit (£000s)	Increase in profit over previous year (%)	Log numbers of the profit figures
1979	1000	100	3.0000
1980	2000	100	3.3010
1981	3000	50	3.4771
1982	4000	$33\frac{1}{3}$	3.6021

Fig 8.19 *natural scale graph*

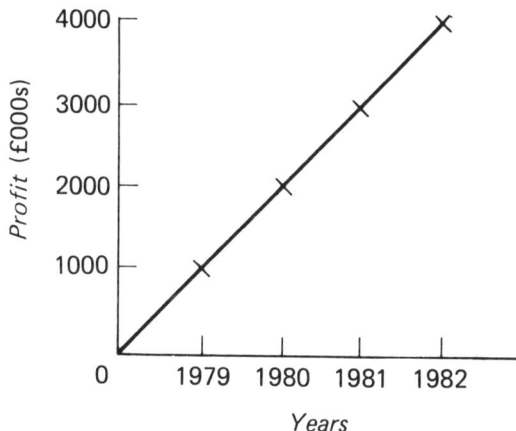

Notice that in Figure 8.20 it is the log numbers that are charted on the vertical axis. This graph shows that although profits are rising, they are rising at a declining rate, while the natural scale graph (Figure 8.19) suggests a constant expansion of profits. They are both 'correct'; they simply show different aspects of the same information.

There are two methods of producing a semi-log or ratio graph:

(i) By plotting the graph on special graph or ratio-scale paper.
(ii) By plotting the log of the data on the vertical axis.

Fig 8.20 *semi-log graph*

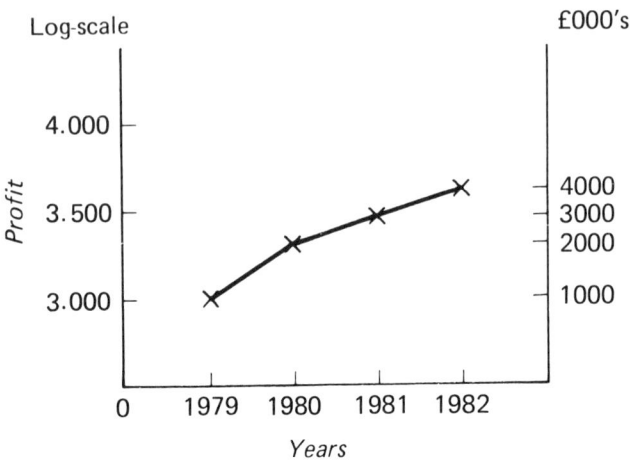

Interpretation of semi-log graphs:

(i) The curve with the greatest slope is the one with the greatest rate of increase.
(ii) The slope of the curve indicates the rate at which the figures are increasing.
(iii) If the 'curve' is a straight line the rate of increase remains constant.
(iv) If the absolute increase is constant the curve will become progressively less steep (as in Figure 8.20).

8.10 STRAIGHT LINE GRAPHS

These occur when there is a particular kind of arithmetical relationship between two sets of data. The relationship is 'direct variation', in which a change in one variable is matched by a similar change in the other variable. *For example*:

(i) Every pound note is worth exactly the same number of dollars and therefore as pounds are converted to dollars the relationship remains in direct variation. If £1 = $2, then £10 = $20 and £100 = $200.
(ii) Hourly paid work involves a direct variation between pay and hours. If the rate of pay is 90 pence an hour, then:
20 hours worked = 20 × 90p = £18
40 hours worked = 40 × 90p = £36
60 hours worked = 60 × 90p = £54

This can be shown on a graph (Figure 8.21):

Fig 8.21 *straight line graph*

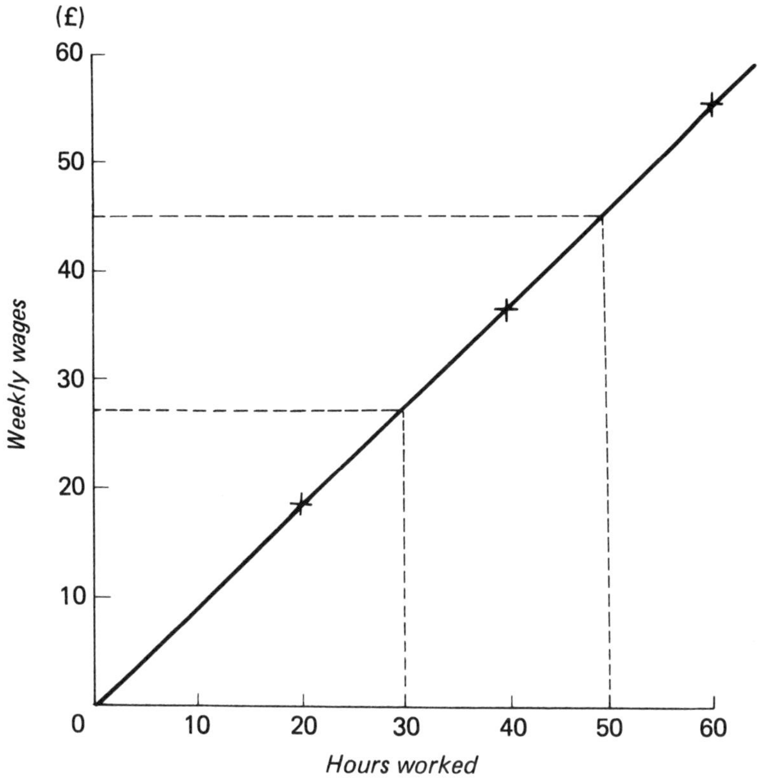

The advantage of drawing the graph is that it is possible to interpolate information from it. In fact the graph can be used as a ready reckoner:

30 hours of work will earn £27,
50 hours of work will earn £45.

The straight line graph is an application of graphical methods. Other pictorial methods are also applied to particular circumstances and uses. The following examples are some of these.

8.11 THE GANTT CHART

This is often used in production as a progress chart. Usually it consists of two horizontal bar charts for each period of time. One bar may indicate the planned production or running time and the other bar the actual figures. Any discrepancy between the two reveals a loss of production.

For example: Figure 8.22

Fig 8.22 *gantt chart*

Production record

	Monday	Tuesday	Wednesday	Thursday	Friday
Planned production					
Actual production					

In Figure 8.22 the horizontal scale shows 100% for each day, which indicates the total possible production. The thinner line shows the planned production based on the variable factors in production. The thicker bar shows the actual production achieved each day during a particular week.

The chart shows that the amount of production planned is much the same on Monday, Tuesday and Wednesday and rather less on Thursday and Friday. Actual production has equalled planned production only on the Monday and fell well short on the Friday.

This type of information may be used as the basis for action, for instance to increase actual output, or to reduce target production to realistic levels. These charts can be used to control the use of machines and in various other aspects of production.

8.12 BREAK-EVEN CHART

This is a chart which shows the profit or loss for any given output. The simplest chart shows two curves or straight lines, one showing the relationship between revenue and output, the other the relationship between cost and output.

For example: a particular make of radio has variable costs of £30 per unit, overall fixed costs of £100,000 and a selling price of £60 a unit. This produces a table of output, costs and revenue (Table 8.3):

Table 8.3 **break-even table**

Units (000s)	Costs (£)	Revenue (£)
1	130,000 (30 + 100,000)	60,000
2	160,000	120,000
3	190,000	180,000
4	220,000	240,000
5	250,000	300,000
6	280,000	360,000

The data in Table 8.3 can be used to construct a break-even chart (Figure 8.23).

Fig 8.23 *break-even chart*

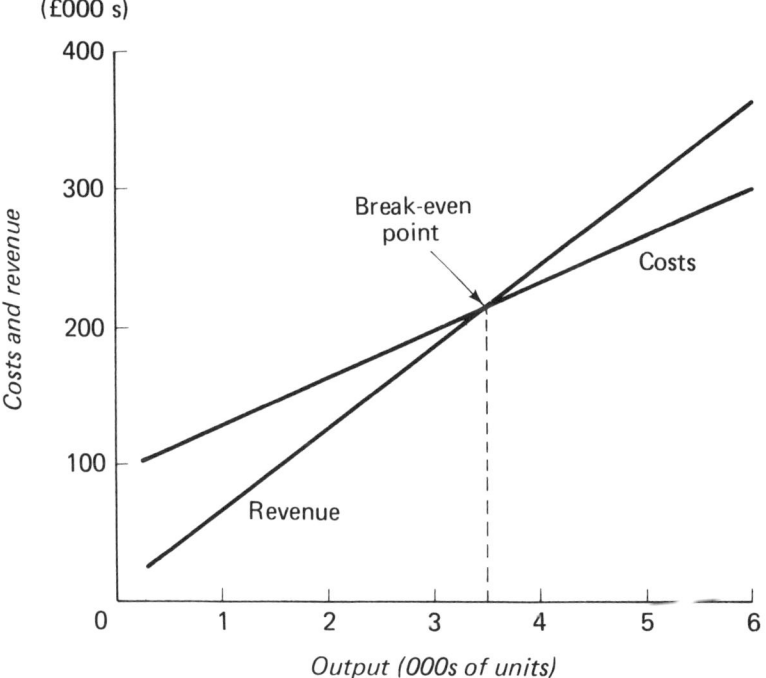

Where the two lines cross is the break-even point (at about 3500 units of output). At this output revenue covers costs, while below this output there is a loss and above this point there is a profit.

8.13 THE Z CHART

The Z chart consists of three graphs plotted together on the same axes. The three graphs are:

(i) the original data
(ii) the cumulative total
(iii) the moving annual total

When plotted on a graph the three curves form the shape of a 'Z'. The curve of the original data shows the current fluctuations, the cumulative curve shows the position to date and the trend is indicated by the moving annual total. Therefore this chart can be used to compare the basic data with trends of data, such as sales.

For example: Table 8.4 and Figure 8.24.

Table 8.4 **monthly sales**

Month	Monthly sales (£)	Cumulative monthly total (£)	Moving annual total (£)
January	1,000	1,000	12,000
February	1,200	2,200	13,000
March	800	3,000	13,200
April	1,300	4,300	14,000
May	1,450	5,750	14,300
June	2,000	7,750	15,000
July	2,150	9,900	15,400
August	2,000	11,900	16,400
September	1,500	13,400	16,500
October	1,100	14,500	15,200
November	800	15,300	15,600
December	700	16,000	16,000

The moving annual total is obtained by adding a new month each time and dropping a month from the previous year. To calculate it requires the figures for two years (see Chapter 14).

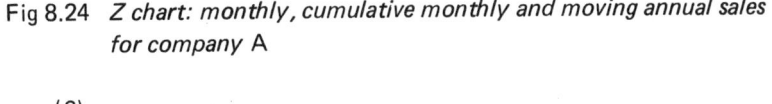

Fig 8.24 *Z chart: monthly, cumulative monthly and moving annual sales for company A*

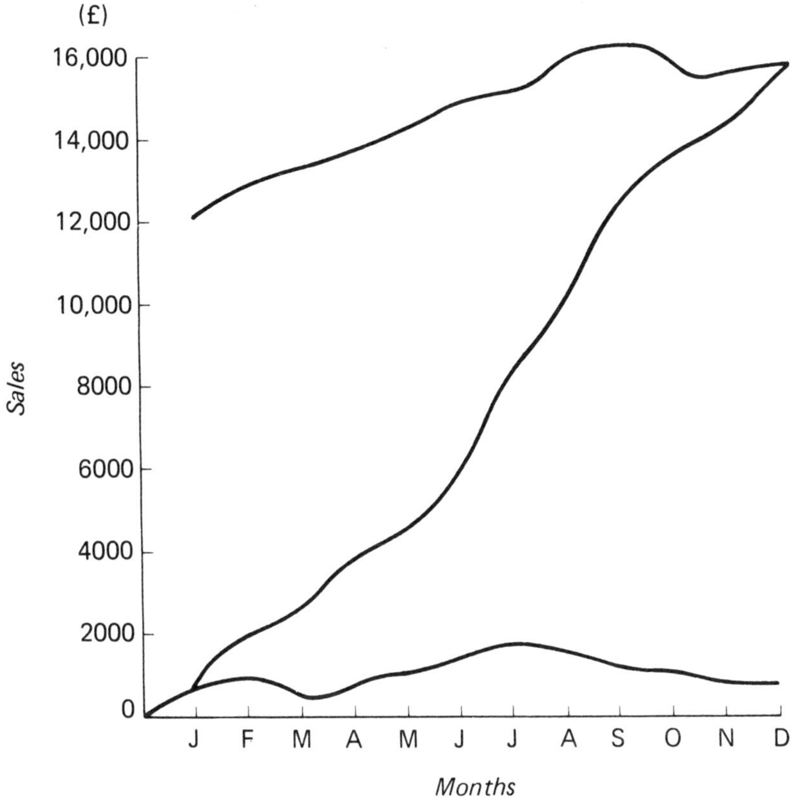

8.14 THE LORENZ CURVE

This is a graphical method of showing the deviation from the average of a group of data. It is a cumulative percentage curve.

The curve is often used to show the level of inequality. For instance, it can show the number of people saving against the amount saved. The more equal the distribution of saving, the flatter the curve. If there was equality between the two variables, the curve would be a straight line, equal to the 'line of equal distribution'.

The Lorenz curve gives an immediate impression and it is used for comparison rather than as a quantitative measure of inequality.

For example: Table 8.5 and Figure 8.25 show comparison between the number of people and accumulated wealth.

Table 8.5 **accumulated wealth**

Income (£000s)	No. of people (f)	Accumulated wealth (£000)	Accumulative wealth (£00s)	(%)	Cumulative frequency (f)	(%)
Less than 5	144	32	32	16	144	48
5-9.9	54	22	54	27	198	66
10-14.9	36	24	78	39	234	78
15-19.9	24	20	98	49	258	86
20-24.9	18	24	122	61	276	92
25-29.9	15	26	148	74	291	97
30-34.9	9	52	200	100	300	100

Fig 8.25 *Lorenz curve*

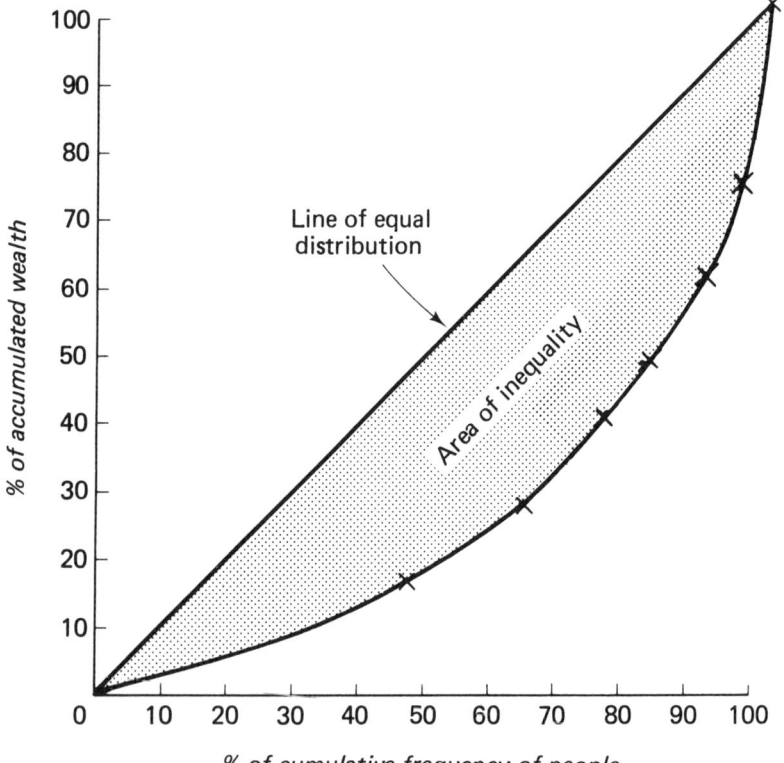

Figure 8.25 indicates that the distribution of wealth is not very equal. If there was 'equality' then 10% of the income earners would have 10% of the total wealth and 50% of the earners would have 50% of the total wealth and so on.

8.15 PRESENTATION AND PERCEPTION

Charts and graphs can illustrate information, emphasise the central points as well as areas of difference and similarity, and clarify the meaning of complicated data.

Presentation involves a process of summarisation and simplification. One problem with this is that it can produce over-simplification and also it is possible to present information in such a way that a wrong impression is given.

The problems of presentation can be said to include:

(a) Difficulties of perception.
(b) Problems of distortion and deception.

(a) **Perception**: this can cause problems because it is not always possible to be sure what is seen.

For example:

(i) In Figure 8.26, which segment of the line *ABC* is the greater: *AB* or *BC*?

Fig 8.26 ABC *line*

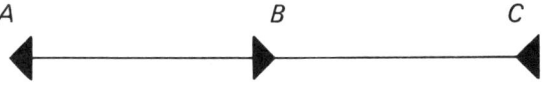

Fig 8.27 BX *and* AC *lines:*

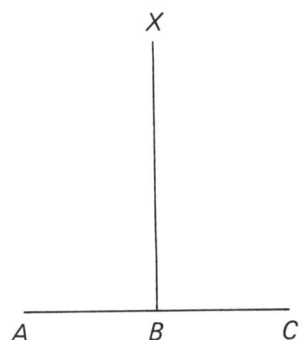

When the line in Figure 8.26 is measured, it will be seen that AB and BC are exactly the same length.

(ii) Which is longer in Figure 8.27, BX or AC?

In fact, again the lines are the same length.

(iii) Then there is the famous young woman, old woman (or wife and mother-in-law) ambiguity: Is Figure 8.28 a picture of a young woman or of an old woman?

Fig 8.28 *young woman, old woman*

In experiments 60% of people have seen the younger woman, 40% the older woman. In fact both women are in the picture.

These three examples highlight the point that it is not possible to take the perception of illustrations and diagrams for granted.

(b) **Distortion and deception**: these may occur with any illustrations and involve the misrepresentation of information.

Examples include bar charts with bars of different width, pictograms with different size symbols which are not in proportion, and graphs which do not keep to the basic rules.

Graphs may be distorted by:

(i) not starting the y axis at zero.

The distortion in Figure 8.30 makes the curve appear steeper than it does in Figure 8.29 where the origin is shown; this gives the impression that sales have increased more rapidly than they have in fact.

(ii) compressing the vertical or horizontal axis (Figures 8.31 and 8.32).

Fig 8.29 *actual sales*

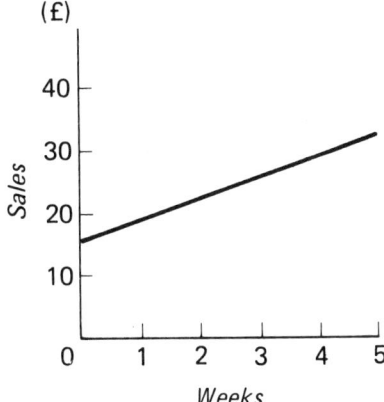

Fig 8.30 *sales not showing the zero*

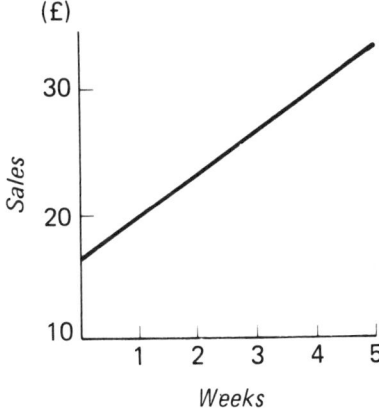

Fig 8.31 *compressed vertical axis*

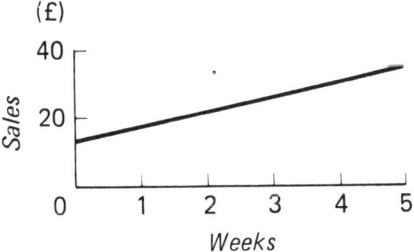

Fig 8.32 *compressed horizontal axis*

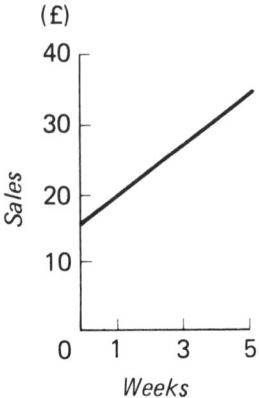

These graphs flatten or sharpen the angle of the curve.

All forms of distortion and deception or illustrations which encourage ambiguity of perception work against the basic aims of statistical presentation, which are clarity of communication and accuracy in the presentation of information.

ASSIGNMENTS

1 Look in any government publication and select three different types of statistical presentation. Discuss the reasons why each particular type of presentation was chosen and whether other forms of presentation might have been just as good.

2 Draw suitable diagrams and graphs to illustrate Table 1.1. Briefly justify your choice of diagram.

3 Discuss the problems of perception and distortion in the presentation of information. Comment on particular advertisements which make use of these 'concepts'.

4 Carefully and fully interpret the information shown in Figures 8.1, 8.2, 8.3, 8.4, 8.5, 8.6, 8.8, 8.9.

5 Discuss the advantages of the various forms of graph compared with histograms and bar charts.

6 Look at the annual report of a company and discuss critically the forms of statistical presentation used.

CHAPTER 9

SUMMARISING DATA:

AVERAGES

9.1 THE ROLE OF THE AVERAGE

Averages are measures of central tendency and measures of location. As a measure of central tendency an average provides a value around which a set of data is located. As a measure of location an average provides an indication of whereabouts the data is situated. An average price can give an indication of whether a particular commodity is likely to cost £1 or £10 or £100.

An average summarises a group of figures and represents it in the sense that the average provides an immediate idea about the group. An average can provide a description of a group of items so as to distinguish it from another group with similar characteristics.

Therefore an average is a way of describing data very concisely, and because of this there are a number of averages which can be used depending on the type of description required. The three most commonly used averages are the arithmetic mean, the median and the mode.

Averages are used all the time in everyday life and work. The statement that 'incomes have risen by 10% in the last year' is usually qualified by the word 'average': 'average incomes have risen by 10% in the last year'. The *average* consumer is said to buy more each year, the *average* rainfall has risen or fallen, the *average* height of people . . . , the *average* price of electrical goods . . . and so on.

In general averages can be said to:

(1) summarise a group of figures, smoothing out abnormalities in a way that is useful for comparison.

For example: in cricket, if a batsman has scored 1000 runs in 40 innings in all of which he is out, the scores in each innings may have varied between 0 and 200. The average would be 25 runs per innings (1000 divided by 40). This figure can be compared with the average achieved by other batsman.

Another example: two firms may pay very widely varying wages to similar types of employees. In one firm, wages may vary between £45 a week and £125; in the other firm, wages may vary between £85 and £150. Both firms may have an average wage for this type of employee of £100.

(ii) An average can provide a mental picture of the distribution it represents.

For example: it may be recorded that a shop that is for sale has average weekly sales of £5000. This provides an immediate idea of the size of the business, although sales might have been £10,000 in Christmas week and £500 in the worst week.

(iii) An average can provide valuable knowledge about the whole distribution.

For example: if the average wage in a factory is £100 and there are 3000 employees, then it can be deduced that the weekly wage bill is £300,000.

(iv) The word 'average' is used in daily conversation and is used loosely.

For example: the statement 'I think that on average I use about ten gallons of petrol a week', is using the average as an estimate.

(v) Averages can conceal important facts.

For example: two companies may both have average annual profits of £50,000 over the last five years. However, when their actual records are inspected it may be found that the companies' performances are very different:

	1978	1979	1980	1981	1982
	£	£	£	£	£
Company A	110,000	80,000	35,000	20,000	5,000
Company B	5,000	20,000	35,000	80,000	110,000

In the same way, it is important to know not only the average but also other figures such as the minimum and maximum. An engineer designing a reservoir must know not only the average rainfall of the region, but also the maximum.

(vi) Therefore averages can provide the first stage of an investigation, but do not provide all the information required for many purposes.

9.2 THE ARITHMETIC MEAN

This is the average to which most people refer when they use the word 'average'.

The arithmetic average can be defined as the sum of the items divided by the number of them.

For example: if five people have £15, £17, £18, £20, £30 respectively, the arithmetic mean is £20. That is, if £100 was shared equally between five people, they would have £20 each.

$$\text{The arithmetic mean} = \frac{\text{total value of items}}{\text{total number of items}}$$

In the above example the arithmetic mean

$$= \frac{£100}{5} = \underline{£20}$$

$$\text{The arithmetic mean} = \frac{\Sigma x}{n}$$

where Σ = the sum of

x = the value of the items
n = the number of items

The usual symbol for the arithmetic mean is $\bar{x}$ (x-bar or bar-x). Strictly, $\bar{x}$ is the symbol for the the arithmetic mean of a sample and μ (mu or mew) is the symbol for the arithmetic mean of the population from which samples are selected. However, $\bar{x}$ is used very widely.

Another method of calculating the mean is to assume an average by inspection, find the deviation of the items from this assumed average, sum the deviations, average them and add or subtract this from the assumed average.

For example: if five people have £15, £17, £18, £20, £30, the average can be assumed to be £18. Deviations from this assumed mean will be:

Item (£)	Deviations
15	−3
17	−1
18	0
20	+2
30	+12
	−4 + 14 = +10

$\bar{x} = 18 + \frac{10}{5}$
$= 18 + 2$
$= \underline{£20}$

Notice that the formula used here is:

$$\bar{x} = x \pm \frac{\Sigma x}{n}$$

where x represents the assumed mean. In this case the assumed mean is £18. The deviations of the items from this assumed mean total +10. This is a total deviation of +10 over 5 items, averaging 2 per item. The 2 is added (in this case) to the assumed mean, to arrive at the same answer (£20) as the first method.

With large numbers of figures this method may be faster than the first method. Also it is useful to have an introduction to it, for further calculations.

Any figure chosen as the assumed mean would give the correct answer.

9.3 THE ARITHMETIC MEAN OF A FREQUENCY DISTRIBUTION

This can be described also as the weighted arithmetic mean, or the arithmetic mean of grouped data.

This is calculated by multiplying the item by the frequency or weight, adding them up and dividing by the sum of the frequencies.

The formula is:

$$\bar{x} = \frac{\Sigma fx}{\Sigma f}$$

where Σ = the sum of
f = the frequency
x = the value of the items

With the use of the assumed mean this becomes:

$$\bar{x} = \frac{\Sigma fdx}{\Sigma f}$$

where x = the assumed mean
d_x = deviation from the assumed mean
fd_x = the frequency times the deviation from the assumed mean

For example: Table 9.1.

Table 9.1 **the arithmetic mean of a frequency distribution**

Price (to the nearest £)	Number of transistor radios (f)	Deviation from assumed mean (d_x) ($x = 28$)	Frequency x deviation from assumed mean (fd_x)
20	2	−8	−16
24	6	−4	−24
25	10	−3	−30
30	4	+2	+8
32	3	+4	+12
	25		−70 +20
	$\Sigma f = 25$		$\Sigma fd_x = -50$

$$\bar{x} = x \pm \frac{\Sigma fd_x}{\Sigma f}$$
$$= 28 - \frac{50}{25}$$
$$= 28 - 2$$
$$= \underline{£26}$$

This shows that the average price of these 25 transistor radios is £26. The method used to calculate the average was by using the assumed mean. This result can be checked by using the other method of calculation. This is to multiply the items by the frequencies and to divide the sum by the total frequencies (see Table 9.2 and the calculations that follow).

Table 9.2 **price of transistor radios**

Price (£)	Number	Price x frequency
20	2	40
24	6	144
25	10	250
30	4	120
32	3	96
	$\Sigma f = 25$	$\Sigma f_x = 650$

$$\bar{x} = \frac{\Sigma fx}{\Sigma f}$$
$$= \frac{650}{25}$$
$$= \underline{£26}$$

This method helps to emphasise the fact that the distribution shows that although there are some transistor radios priced at over £30 and others at £20, the average price for this selection is £26. If this was a random sample of all transistor radios then the consumer would know that £26 was likely to be a representative price for this type of radio (see Chapter 11).

9.4 THE ARITHMETIC MEAN OF A GROUPED FREQUENCY DISTRIBUTION

This can also be described as the weighted arithmetic mean with frequency classes. The problem with class intervals is that there is no way of knowing the actual distribution of frequencies within a class. Therefore an assumption has to be made and this is usually that the frequencies equal the mid-point of the class interval (see Section 7.3).

There are two methods of calculating the arithmetic mean of a frequency distribution with class intervals:

(a) The mid-point method.
(b) The class interval method.

(a) The mid-point method

This method uses the mid-points of each class to represent the classes in the calculation. The mid-point is usually found by adding together the lower and upper limits of the class and dividing by 2. With continuous data it can be assumed that class intervals such as '20 but less than 25' in fact mean '20 to 24.99'. The next class interval will be '25 to 29.99' and so on.

Rounded to one decimal place this mid-point will be 22.5:

$$\frac{20 + 24.99}{2} = \frac{44.99}{2} = 22.49$$

It could be said that 22.5 is the 'common-sense' mid-point, in the sense that it does not suggest a higher degree of accuracy than is involved in the whole calculation.

Therefore for class intervals such as '20 but less than 25' the mid-point can be found by:

$$\frac{20+25}{2} = \frac{45}{2} = 22.5$$

With a class interval of '20 to 24 followed by 25 to 29, 30 to 34 and so on' the mid-point can be found by:

$$\frac{20+24+1}{2} = \frac{45}{2} = 22.5$$

Instead of adding 1, it is possible to add the lower limits of adjacent class intervals, 20 + 25 (see Section 7.3 for a discussion of the problems of classification).

For example: Table 9.3 and following calculations.

Table 9.3 **the mid-point method**

Overtime pay (£)	Number of employees	Mid-points of class intervals	Deviation of mid-point from assumed mean	Frequency × d_x	
	(f)	$(m-p)$	(d_x)	(fd_x)	
20 but less than 25	11	22.5	−15	−165	
25 but less than 30	15	27.5	−10	−150	
30 but less than 35	16	32.5	−5	−80	
35 but less than 40	18	37.5	0	0	
40 but less than 45	30	42.5	+5		+150
45 but less than 50	10	47.5	+10		+100
	100			−395	+250

$\Sigma f = 100$ $\Sigma fd_x = -145$

$$\bar{x} = x \pm \frac{\Sigma fdx}{\Sigma f}$$

$= £37.5 - \frac{145}{100}$

$= £37.5 - 1.45$

$= \underline{£36.05}$

(b) The class-interval method

In this case instead of using the mid-points of the class intervals, the calculation is carried out in 'units of the class interval'. In table 9.4 the deviation of classes from the assumed mean are in units of 5, and this has to be allowed for in the calculation by multiplying by 5.

Table 9.4 the class-interval method

Overtime pay (£)	Number of employees (f)	Deviation of classes from class of the assumed mean (d_x)	Frequency × d_x (fd_x)
20 but less than 25	11	−3	−33
25 but less than 30	15	−2	−30
30 but less than 35	16	−1	−16
35 but less than 40	18	0	0
40 but less than 45	30	+1	+30
45 but less than 50	10	+2	+20
	100		−79 +50

x = £37.5 $\qquad\qquad\qquad\qquad\qquad\qquad\qquad\qquad$ $\Sigma fd_x = -29$

$$\bar{x} = x \pm \frac{\Sigma fdx}{\Sigma f} \times \text{class interval}$$
$$= £37.5 - \frac{29}{100} \times 5$$
$$= £37.5 - 0.29 \times 5$$
$$= £37.5 - 1.45$$
$$= \underline{£36.05}$$

The class-interval method is very often the simpler method for calculation. However, if the class intervals are uneven allowance must be made for this.

For example: if the last class in Table 9.4 had been '45 but less than 55', then there would have been two units of 5 in this class. The deviation of this class from the class of the assumed mean would have to have been +2.5. If the mid-point of this larger class is taken as 50, then this is 2.5 units of 5 away from the assumed mean of 37.5.

With a number of unequal class intervals it may be easier to use the mid-point method.

9.5 ADVANTAGES AND DISADVANTAGES OF THE ARITHMETIC MEAN

(a) **the advantages of the arithmetic mean**
  (i) It is widely understood and the basic calculation is straightforward.
  (ii) It makes use of all the data in the group and it can be determined with mathematical precision.
  (iii) It can be determined when only the total value and the number of items are known.

(b) **the disadvantages of the arithmetic mean**
  (i) A few items of a very high or very low value may make the mean appear unrepresentative of the distribution.
  (ii) It may not correspond to an actual value and this may make it appear unrealistic.
  (iii) When there are open-ended class intervals, assumptions have to be made which may not be accurate.

The arithmetic mean does involve the use of all the data and all the values and this major strength is also a weakness under certain circumstances. The arithmetic mean of 2, 5, 6, 8 and 129 is 30. This answer is not close to any of the actual values. However, it is because all the values are used, including the extreme ones, and because it is capable of precise calculation that statisticians prefer the arithmetic mean to other averages for most purposes.

9.6 THE MEDIAN

The median can be defined as the value of the middle item of a distribution which is set out in order.

For example: if five people earn
    £60, £70, £100, £115 and £320,
The median is the value of the middle item: £100.

In a discrete distribution such as this one, the median can be ascertained by inspection.

The formula for finding the median position for a discrete series is:

$$\text{The median} = \frac{n+1}{2}$$
$$= \frac{5+1}{2}$$
$$= 3$$

The median is the value of the third item, or £100.

If there are an even number of items, the two middle items are added together and divided by 2.

For example: if six people earn

£60, £70, £100, £110, £115, £320

$$\text{The median position} = \frac{n+1}{2} = \frac{6+1}{2} = 3.5$$

$$\text{The median} = \frac{£100 + £110}{2} = \underline{£105}$$

The symbol for the median is M. In this example the position of M is between the third and fourth item and therefore the value of M is £100 plus £110 divided by 2: £105.

This example shows that the median is unaffected by extreme values (such as the £320) which is one reason for using the median as an average in certain circumstances. It is often used with wage distributions.

Also, the median divides a distribution in half by the number of items (not their values). In the examples above, where the median is £100, there are two items on either side, and where the median is £105, there are three items on either side.

With a *grouped frequency distribution* the median can be found by two methods:

(a) by calculation,
(b) graphically.

The position of the median for a continuous series is $\frac{n}{2}$ or $\frac{f}{2}$.

Therefore in Table 9.5 the position of the median

$$= \frac{50}{2} = 25$$

The 25th employee's overtime pay is the median pay. This falls within the class interval £15 but less than £20. Therefore the 25th employee earns at least £15 in overtime pay, but less than £20.

The only way to arrive at a closer calculation of his pay is to assume that the earnings are evenly distributed within the class interval. The 25th employee is the third person in the group, because the previous group includes the 22nd employee and $25 - 22 = 3$.

(a) The median found by calculation

Table 9.5 the median

Overtime pay (£)	Number of employees	Cumulative frequency
0 but less than 5	3	3 people received less than £5
5 but less than 10	5	8 people received less than £10
10 but less than 15	14	22 people received less than £15
15 but less than 20	12	34 people received less than £20
20 but less than 25	10	44 people received less than £25
25 but less than 30	6	50 people received less than £30
	50	

The £5 in the class interval is divided evenly between the 12 people in the group and the 25th employee is the 3rd person of these 12.
Therefore

$$M = £15 + \frac{3}{12} \times 5$$

where 15 is the lower class interval of the class in which the median falls, and $\frac{3}{12} \times 5$ indicates that the median pay is the third out of 12 sharing £5. (Notice that it is only the class interval of the class in which the median falls that is important; therefore irregular class intervals in a distribution are not a problem because they can be ignored.)

Therefore $M = £16.25$

In this distribution the median overtime pay is the earnings of the 25th employee which is £16.25.

This divides the distribution in half, so that of the 50 employees half will earn less than £16.25 and half will earn more. In this sense the median provides a good representation of the grouped frequency distribution.

(b) The median found graphically

By this method the median is found by drawing the cumulative frequency curve or ogive. 'Ogive' is an architectural term which is used to describe an S (or flattened S) shape, similar to the usual shape of the cumulative frequency curve.

For example: Table 9.6 and Figure 9.1.

Table 9.6 **the cumulative frequency**

Overtime pay (£)	Number of employees	Cumulative frequency
0 but less than 5	3	3
5 but less than 10	5	8
10 but less than 15	14	22
15 but less than 20	12	34
20 but less than 25	10	44
25 but less than 30	6	50
	50	

From Table 9.6 the graph in Figure 9.1 can be drawn, plotting the cumulative frequency against overtime pay.

Fig 9.1 *the cumulative frequency curve*

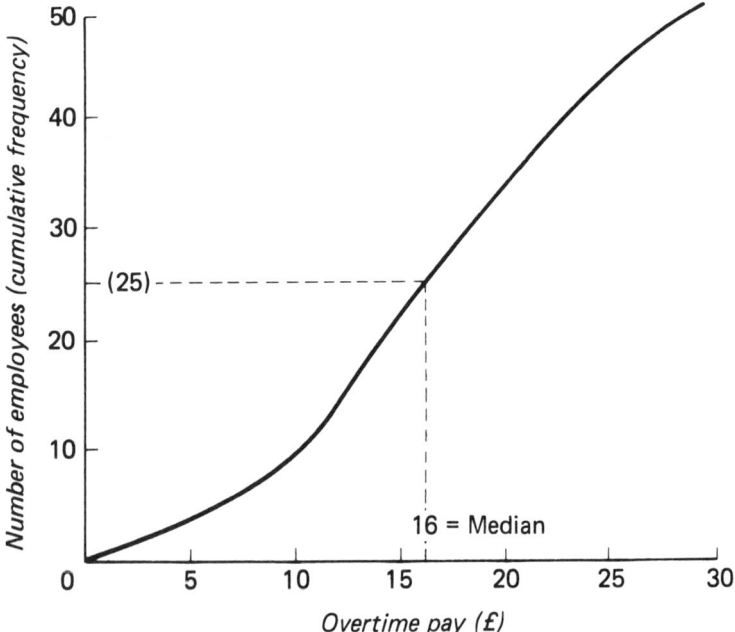

Notice that in Figure 9.1, when plotting the frequencies, the point must be placed at the end of the group intervals (5, 10, 15 etc.)

$$\text{The position of } M = \frac{50}{2} = 25$$

A horizontal line is drawn from this point (25) to the ogive, and at the point of intersection a vertical line is dropped to the horizontal axis. This shows that the median is about £16. The more accurately the graph is drawn, the more accurately it is possible to interpolate the median.

(c) **The advantages of the median**
 (i) Extreme high and low values do not distort it as a representative average. Therefore it is useful for describing distributions in areas such as wages where a few extreme values would distort the arithmetic mean.
 (ii) It is straightforward to calculate even if not all the values are known, or where there are irregular class intervals.
 (iii) It is often an actual value and even when it is not it may 'look' representative and realistic.

(d) **The disadvantages of the median**
 (i) It gives the value of only one item. The other items are important in ascertaining the position of the median, but their values do not influence the value of the median. If the values ar spread erratically, the median may not be a very representative figure.
 (ii) In a continuous series, grouped in class intervals, the value of the median is only an estimate based on the assumption that the values of the items in a class are distributed evenly within the class.
 (iii) It cannot be used to determine the value of all the items; the number of items multiplied by the median will not give the total for the data; therefore it is not suitable for further arithmetical calculations.

9.7 THE QUARTILES

The median divides an ordered distribution into half, and in a similar way it is possible to divide distributions into quarters, tenths and so on. The method is similar in all cases.

The most frequently used division is into quartiles. These divide an ordered distribution into four equal parts.

There are 'three' quartiles:

 (i) The lower quartile: Q_1

(ii) The middle quartile: Q_2 (this is the same as the median: M)
(iii) The upper quartile: Q_3

In a distribution of 100 items the quartiles will be the values of the 25th (Q_1) and the 75th (Q_3) items. With the median, which will be the value of the 50th item, the two quartiles will divide this distribution into four equal parts: 1-24, 26-49, 51-74, 76-99.

The method of calculating the lower and upper quartiles is very similar to the method for calculating the median.

For example:

Table 9.7 **the lower and upper quartiles**

Overtime pay (£)	Number of employees	Cumulative frequency
0 but less than 5	3	3
5 but less than 10	5	8
10 but less than 15	14	22
15 but less than 20	12	34
20 but less than 25	10	44
25 but less than 30	6	50
	50	

The position of the lower quartile (Q_1) = $n/4$. In Table 9.7 this will be 50/4 = 12.5.

The pay received by a theoretical employee lying between the 12th and 13th is the lower quartile pay.

This lies in the class '£10 but less than £15', which is shared by 14 employees.
Therefore

$$Q_1 = £10 + \frac{4.5}{14} \times 5$$
$$= £10 + \frac{22.5}{14}$$
$$= £10 + 1.607$$
$$= \underline{£11.61}$$

The position of the upper quartile (Q_3) = $3n/4$. In Table 9.7 this will be $3 \times 50/4 = 150/4 = 37.5$.

The pay received by a theoretical employee lying between the 37th and 38th employee is the upper quartile pay.

This lies in the class '£20 but less than £25', which is shared by 10 employees.

$$Q_3 = £20 + \frac{3.5}{10} \times 5$$
$$= £20 + \frac{17.5}{10}$$
$$= £20 + 1.75$$
$$= \underline{£21.75}$$

Therefore this distribution is divided in the following way:

Q_1 = 12.5th employee receiving £11.61 in overtime pay
M = 25th employee receiving £16.25 in overtime pay
Q_3 = 37.5th employee receiving £21.75 in overtime pay

Graphically this can be interpolated as shown in Figure 9.2.

Fig 9.2 *the lower and upper quartiles*

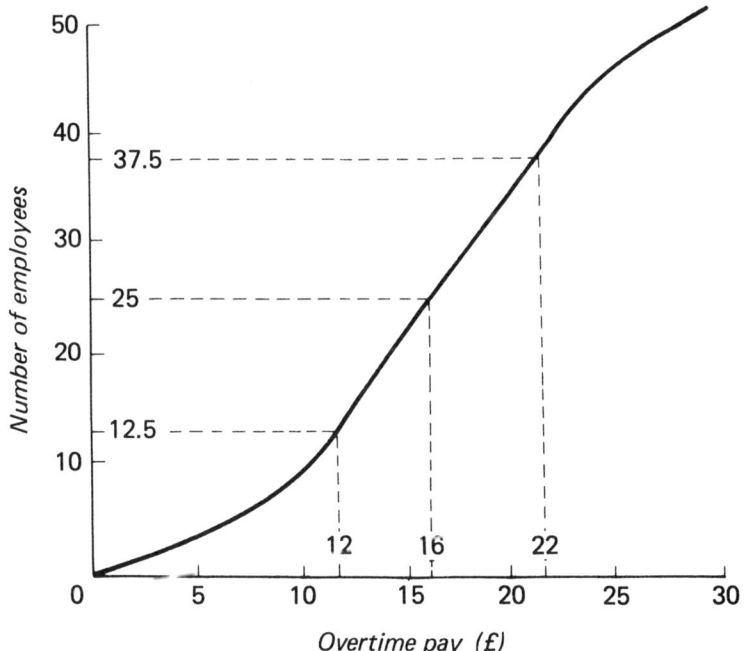

From the division of the distribution in Table 9.7 and Figure 9.2 by the median and the lower and upper quartiles, it can be stated that:

(i) Half the employees are earning less than £16.25 and half more than this in overtime pay.
(ii) Half the employees are earning between £11.61 and £21.75 (that is, between Q_1 and Q_3) in overtime pay.
(iii) A quarter of the employees are earning less than £11.61 and a quarter more than £21.75.

The difference between the upper and lower quartile is called the interquartile range or the quartile deviation (see Section 10.3).

In the same way as the median, the quartiles are useful for descriptive purposes. The results produced from the distribution in Table 9.7 could be compared with those from a similar distribution based on another company.

9.8 THE MODE

This is an average that is frequently used in conversation when reference is made to such things as an 'average' income, an 'average' person and so on. It may be stated that 'the average family has two children', meaning that most families have two children. The arithmetic mean may show that the average family has 2.3 children, but although this may be mathematically correct it may not appear very sensible. In such cases as this, the mode may appear to be a more 'sensible' average to use.

The mode can be defined as *the most frequently occurring value in a distribution*.

Therefore, it is often used in the sense of 'most': 'most' income, 'most' people, 'most' families.

For example: the mode of the figures 4, 4, 5, 6, 11 is 4, because it is the most frequently occurring number.

In a frequency distribution the mode is the item with the highest frequency.

For example:

The size of a component (cm)	The number of makes of car using the size
20	4
21	10
22	15
23	20
24	1

The mode in this example is 23 cm, because 20 is the highest frequency; 23 cm is the modal size of component or the most frequently used size.

Calculating the mode for a grouped frequency distribution is not easy, because since a grouped frequency distribution does not have individual values it is impossible to determine which value occurs most frequently. It is possible to calculate the mode, but it is not particularly useful to do so. For most purposes the modal class is perfectly satisfactory as a form of description.

The modal class is the one with the highest frequency. The problem with using it is that if a different set of class intervals had been chosen when the classifications were being decided, the modal class would have included different values. However, it can be useful as a form of description.

For example: Table 9.8.

Table 9.8 **the modal class**

Overtime pay (£)	Number of employees
0 but less than 5	3
5 but less than 10	5
10 but less than 15	14
15 but less than 20	12
20 but less than 25	10
25 but less than 30	6
	50

In Table 9.8 the modal class is '£10 but less than £15' because this class has the highest frequency (14). If the classification and frequencies had been:

0 but less than 10	8
10 but less than 20	22
20 but less than 30	16

then the modal class would have been '£10 but less than £20'. Although this would have produced a different modal class from the same distribution, the modal class would still have been around the centre because this is a unimodal distribution which is not very skewed (see Section 10.1).

The modal class can be illustrated by drawing a histogram (Figure 9.3):

Fig 9.3 *the modal class*

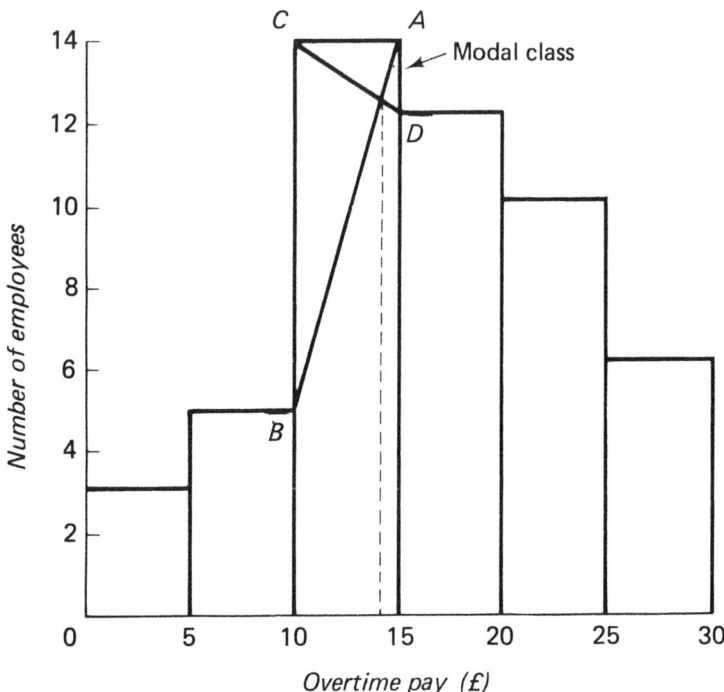

From the histogram in Figure 9.3 it is possible to estimate the mode. This estimation is carried out by drawing a line from the top right-hand corner of the modal class rectangle or block to the point where the top of the next adjacent rectangle to the left meets it (line *A* to *B* in Figure 9.3); and a corresponding line from the left-hand top corner of the modal class rectangle to the top of the class on the right (line *C* to *D*). Where these two lines cross a vertical line can be dropped to the horizontal axis and this will show the value of the mode (about £14 in Figure 9.3). This can only be an estimate, because the individual values are not known. For most purposes, however, the modal class provides a sufficient description of this aspect of a grouped frequency distribution.

Some distributions have more than one modal class, because two or more classes have the same frequency. These distributions are called bi-modal, tri-modal and so on.

For example: Figure 9.4

Fig 9.4 *a bi-modal distribution*

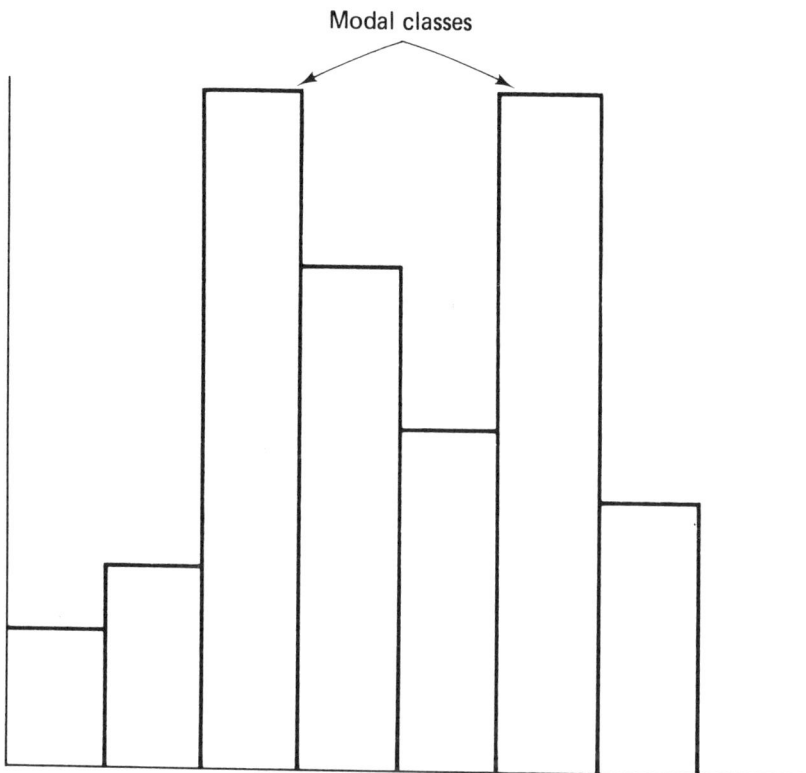

(a) **Advantages of the mode**
 (i) It is a commonly used average, although people do not always realise that they are using it.
 (ii) It can be the best representative of the typical item, because it is the value that occurs most frequently.
 (iii) It has practical uses. For instance, employers will often adopt the modal rates of pay, the rate paid by most other employers. Cars and clothes are made to modal sizes, houses are built on the basis of the modal average size of the family.
 (iv) Modal information can often be supplied quickly by people who have experience in a particular area.
 (v) The mode is often an actual value and therefore may appear to be realistic and 'sensible'.

(b) **Disadvantages of the mode**
  (i) It may not be well defined and can often be a matter of judgement.
  (ii) It does not include all the values in the distribution.
  (iii) It is not very useful if the distribution is widely dispersed.
  (iv) It is unsuitable for further or other kinds of calculation because of its lack of exactness.

9.9 THE GEOMETRIC MEAN

This average is used to measure changes in the rate of growth (see Section 6.14 on the geometric progression).

The geometric mean is defined as *the nth root of the product of the distribution*. If there are three items it is the third root.

For example:

(i) The geometric mean of 3, 4, 15:

$$\sqrt[3]{3 \times 4 \times 15} = \underline{5.6}$$

(ii) If the price of commodity A has risen from £60 to £120, this is an increase of 10%.

If the price of commodity B has risen from £80 to £100, this is an increase of 25%.

The arithmetic mean of these percentage increases would be:

$$\frac{100\% + 25\%}{2} = \underline{62.5\%}$$

The geometric mean would be:

$$\sqrt{100 \times 25} = \underline{50\%}$$

Whether the 'true' increase is 62.5% or 50% is open to argument (see Section 12.1 for a discussion on this in relation to index numbers), however the rate of growth can be described in other terms and the geometric mean is seldom used.

9.10 THE HARMONIC MEAN

The harmonic mean is used to average rates rather than simple values (averaging kilometres per hour, for example). It is rarely used except in engineering.

ASSIGNMENTS

1 *Profits of three companies*

	Company A (£)	Company B (£)	Company C (£)
1976	2,000	15,000	10,000
1977	4,000	2,000	9,000
1978	6,000	6,000	8,000
1979	9,000	2,000	6,000
1980	14,000	14,000	2,000

Calculate the annual average (arithmetic mean) profit for these three companies.

Comment on the results and on the profit figures.

What advice would be reasonable to give to someone thinking of buying one of these companies?

2 *Median and quartiles*

From the following data:

(i) Calculate the median and quartiles.
(ii) Draw an ogive and interpret the median and quartile. Comment on the wage distribution of this company.

A company employing 100 part-time employees makes the following monthly payments:

Part-time pay (£)	Number of people	Cumulative totals
20 and less than 25	5	5
25 and less than 30	8	13
30 and less than 35	17	30
35 and less than 40	39	69
40 and less than 45	14	83
45 and less than 50	12	95
50 and less than 55	5	100
	100	

3 In a factory the operation of 4 machines is recorded. The results are as follows:

Units of output

Machine	Day 1	Day 2	Day 3	Day 4	Day 5
A	500	501	505	504	494
B	560	558	572	560	562
C	530	475	538	442	520
D	460	458	452	462	430

The modal output for all the machines of this type is 500 units a day.
Comment on the performance of these four machines.

4 Discuss the purpose of calculating averages. Why is there more than one type of average?

5 Find references to the arithmetic mean and the median in government publications. Comment on the way these averages are used in these cases.

CHAPTER 10

SUMMARISING DATA: DISPERSION

10.1 DISPERSION

Data can be summarised and compared by the use of averages because they can represent a distribution and provide an indication of location. Also, data can be summarised and compared by measures of dispersion, which are also described as measures of deviation or spread.

If items are widely dispersed, averages do not provide a clear summary of the distribution; they do not give an indication of the form or shape of a distribution. Distributions are not only clustered around a central point, but also spread out around it.

For example: Figure 10.1

The distribution shown in Figure 10.1 is of a bell-shaped or normal curve type (see Section 11.3 for more on the normal curve). It is symmetrical in that the distribution is spread equally on either side of the arithmetic mean, median and mode.

Some frequency distributions are 'skewed', so that the peak is displaced to the left or right (Figures 10.2 and 10.3):
When the peak is displaced to the left of centre, the distribution is described as being positively skewed; when the peak is displaced to the right, the distribution is described as negatively skewed. With these distributions the mode is located at the highest point while the median usually lies between the mode and the mean.

Other commonly occurring shapes of distributions are the bi-modal, the rectangular and the *J*-shaped (Figure 10.4, 10.5 and 10.6).

As a method of providing a short description of distributions of data an average may not be sufficient because it provides an indication of the central value and no more. In many circumstances what is needed is a measure which provides an indication of the deviation of the data around this central value.

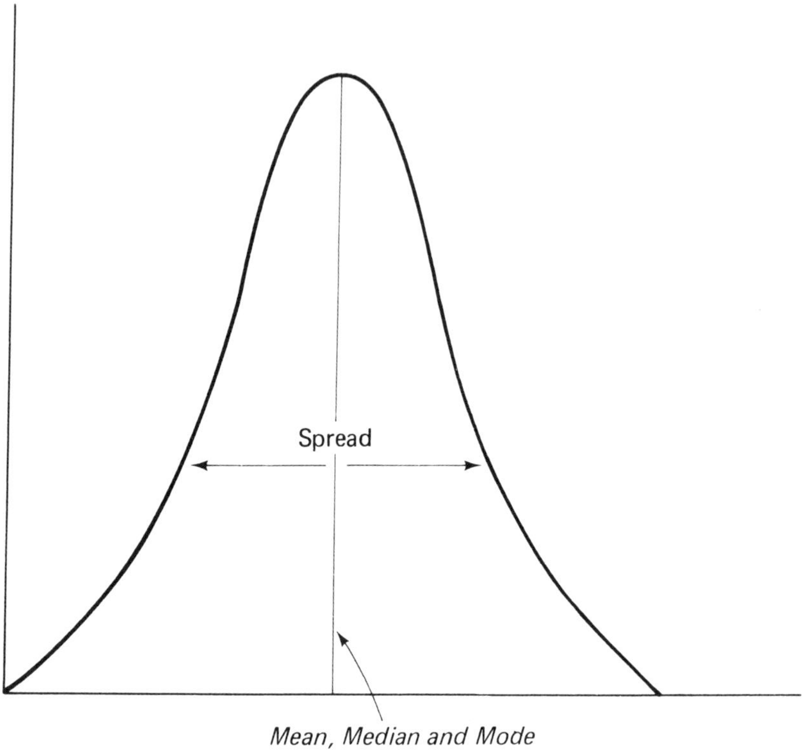

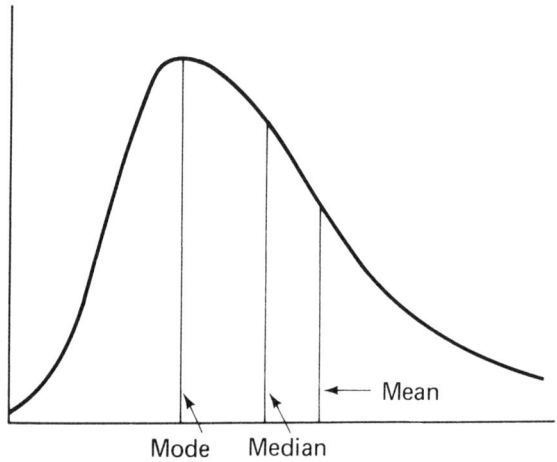

Fig 10.1 *spread*

Fig 10.2 *positively skewed*

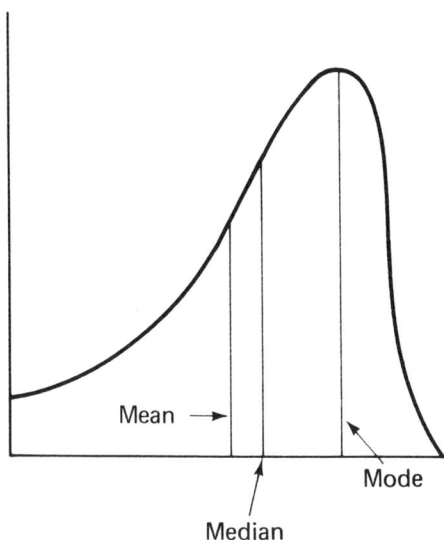

Fig 10.3 *negatively skewed*

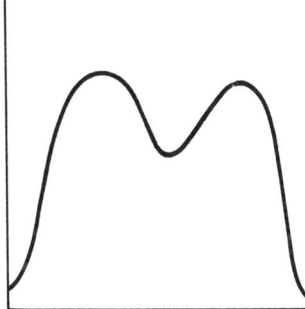

Fig 10.4 *a bi-modal distribution*

Fig 10.5 *a rectangular distribution*

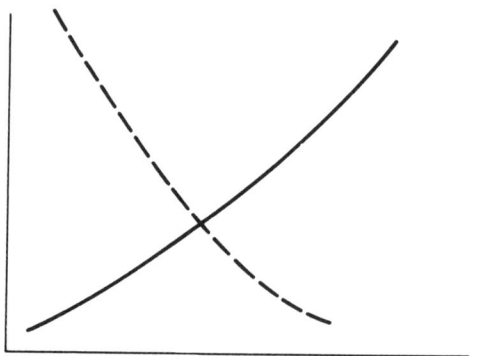

Fig 10.6 *a J-shaped and reverse J-shaped distribution*

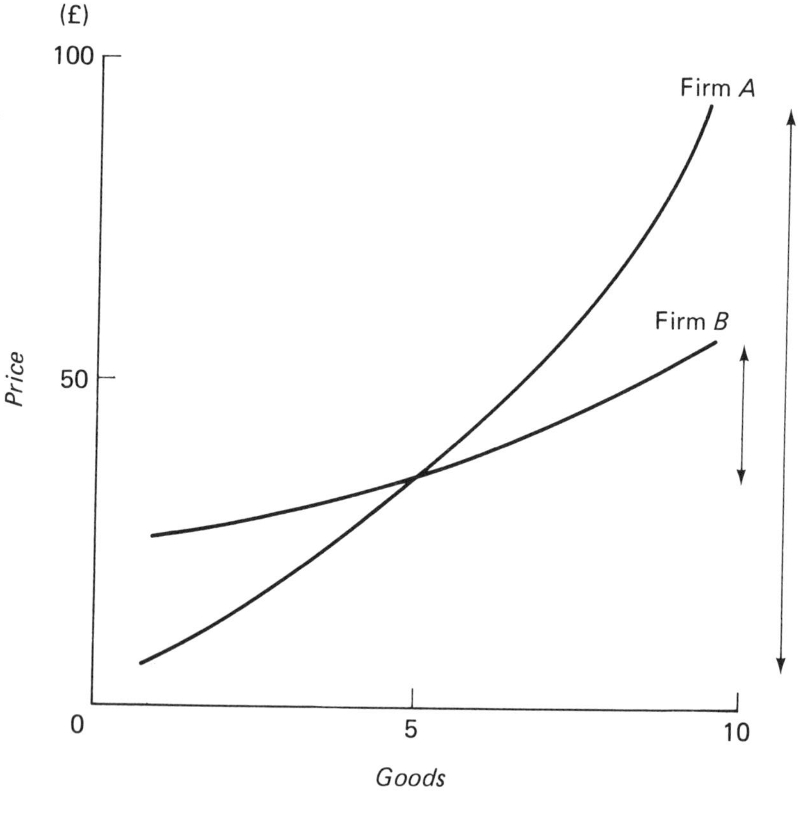

Fig 10.7 *price distribution*

For example: for both the following sets of data the arithmetic mean is 32:

(A) 3, 4, 7, 16, 20, 30, 38, 53, 61, 88.
(B) 20, 21, 23, 28, 31, 34, 37, 39, 49, 50.

In both cases

$$\bar{x} = \frac{320}{10} = 32$$

If a firm sells a variety of makes of a product at an average price of £32, this does not give an indication of the price of the cheapest make it sells or the price of the most expensive. In other words, it does not give an idea of dispersion. If the sets of scores above represent the prices charged by two firms for makes of a particular product they have in stock, then although their average price is the same (£32), the first firm carried very much cheaper (£3) and very much more expensive (£88) makes or brands of this product than the second firm (£20 and £50).

In other words, Firm A has a wider price range (or its prices are more dispersed or more spread out) for this particular product than the price range for Firm B. This can be further illustrated by a graph (Figure 10.7):

It is possible to calculate a measure of spread for these two distributions by using the simplest form of dispersion, the range.

10.2 THE RANGE

The range is an everyday method of describing the dispersion or spread of data.

The range can be defined as the highest value in a distribution minus the lowest.

For example: in the distributions in Figure 10.7 the ranges are:

Firm A, the range = £88 − £3 = £85
Firm B, the range = £50 − £20 = £30

Therefore it is possible to say that Firm A has makes or brands of this product which have an average price of £32 with a range of £85, while Firm B has makes or brands of the same product with an average price of £32 and a range of £30. The larger the range the wider the dispersion or spread of the distribution. Therefore Firm A has a wider spread of prices than Firm B.

In fact the range is often used with the mode. In a shop it is possible to ask (about, for example, watches, cameras or radios): 'What price range do you have?' The answer might be: 'They range from a price of £2 to £200.' This is one simple statement which provides the dispersion or spread of prices for this commodity. The next question could be: 'What is the average price?' After all, the fact that the watches cost between £2 and £200, does not give any indication whether most of them are priced close to £2 or close to £200, or spread evenly through the range. The answer might be: 'the usual price is about £1,' or 'most of our customers buy the ones costing around £10,' or 'the £10 ones are the most popular.' Notice that all three of these answers use the mode rather than the arithmetic mean.

It is now clear that the price of these commodities ranges from £2 to £200 with a modal price of about £10. This means that a customer who buys a watch for £100 is a rare customer buying an expensive watch. The customer who buys a watch for £2 will be buying a very cheap one.

Therefore the range with an average can provide a very clear summary of the variations in the prices (or weights, lengths or sizes) of goods available. This summary does not depend on the number of articles. In the example above, there could have been 20 watches or 200 or 2000.

All measures of dispersion are designed to provide similar information to the range, but information which is either more precise or expressed in a different way.

10.3 THE INTERQUARTILE RANGE

As was seen in Section 9.7 a distribution can be divided into four equal parts and the point marking each quarter is called a quartile. The interquartile range (or quartile deviation) can be defined as the difference between the upper quartile and the lower quartile ($Q_3 - Q_1$).

Table 10.1 shows the wage list of a company employing twenty people. The wages of employees 5, 10 and 15 on this list divide the distribution into four equal parts (1-5, 6-10, 11-15, 16-20). The position and value of the quartiles are:

$$Q_1 = \frac{n}{4} = \frac{20}{4} = 5$$

This is the position of the lower quartile and the wage of the fifth employee is the lower quartile wage: £50

$$Q_3 = \frac{3n}{4} = \frac{60}{4} = 15$$

Table 10.1 **The interquartile range**

Employees	Weekly wage (£)	Employees	Weekly wage (£)
1	40	11	89
2	42	12	93
3	43	13	97
4	48	14	100
5	50	15	110
6	60	16	140
7	62	17	200
8	65	18	210
9	71	19	212
10	80	20	220

This is the position of the upper quartile and the wage of the fifteenth employee is the upper quartile wage: £110

The interquartile range = $Q_3 - Q_1$
£110 − £50 = £60

It is also possible to arrive at the semi-interquartile range by dividing the interquartile range by two:

$$\text{The semi-interquartile range} = \frac{Q_3 - Q_1}{2}$$

$$\frac{£110 - £50}{2} = \frac{£60}{2} = £30$$

The interquartile range helps to summarise and clarify the distribution. In Table 10.1 the interquartile range (£110 minus £50) covers those employees earning the middle range of wages, and this includes 50% of the employees working in the company.

The interquartile range of one company can be compared with that of another. For instance, a company with an interquartile range of wages of £60 has a more widely dispersed middle range of wages than a company with a interquartile range of £20.

Another method of arriving at the value of the interquartile range is by using the ogive (see Section 9.7).

For example: Table 10.2 and Figure 10.8

Table 10.2 **overtime pay**

Overtime pay per week (£)	Number of employees (£)	Cumulative frequency
0 but less than 5	3	3
5 but less than 10	5	8
10 but less than 15	14	22
15 but less than 20	12	34
20 but less than 25	10	44
25 but less than 30	6	50
	50	

Fig 10.8 *interquartile range*

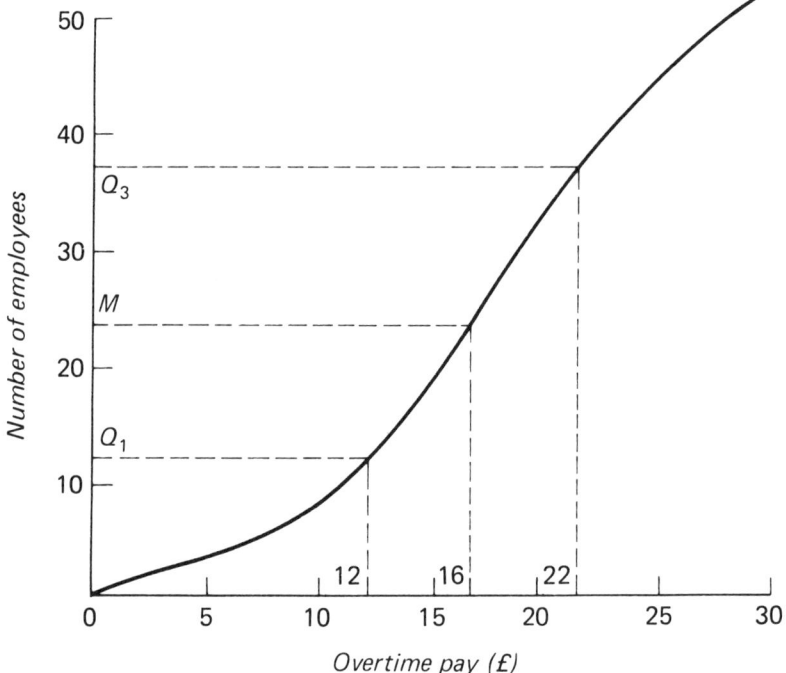

$Q_1 = £12$
$M = £16$
$Q_3 = £22$ $Q_3 - Q_1 = £22 - £12 = \underline{£10}$

The interquartile range is £10 in Figure 10.8, and therefore the middle 50% of the employees of the firm earn a median overtime pay of £16 with an interquartile dispersion of £10.

This result could be compared with another firm with a similar median overtime pay amongst its employees but with an interquartile range of £20. This indicates that in this second company the middle range of employees could earn both less and more overtime pay than the middle range of employees in the first company. Instead of the interquartile range being £22 − £12, it could be say £27 − £7.

Notice that the interquartile range is not influenced by extreme items; therefore very small and very large values do not alter its general spread. For example, if the owner of a company draws an annual salary of £100,000 and the works manager receives an annual salary of £20,000 while the rest of the company employees receive wages between £2000 and £6000 per year, the very high wages will not 'distort' the interquartile range. The 'simple' range would include the extreme figures (range: £2000-£100,000) and the result does not give a very clear picture of the general level of pay in the company, therefore in these circumstances the interquartile range is preferable.

10.4 THE STANDARD DEVIATION

(a) The standard deviation as a measure of dispersion

Neither the range nor the interquartile range make use of all the scores or values in a distribution. The range relies on the highest and lowest values, the interquartile range on the values of the upper and lower quartiles. Therefore both of them rely on the position of two points in a distribution and the values of the other items are not very important. This can lead to 'distortion' in the sense that it is possible that the two values are not very representative of all the other values in the distribution.

The standard deviation is a measure of dispersion which uses all the values in a distribution in the sense that every value contributes to the final result in the same way that every value contributes to the calculation of the arithmetic mean (see Section 9.2).

The standard deviation is the 'standard' measure of dispersion because it is very useful both practically and mathematically. It is important because of its mathematical properties and use in sampling theory (see Chapter 11) rather than because it makes a distribution more easily understood. Therefore it is basically different to the range and the interquartile range which can quickly make a distribution more easily understood but which are of limited use mathematically.

The standard deviation can be used as a measure of dispersion in all symmetrical and unimodal distributions, and also in distributions that are moderately skewed (see Section 10.1). These forms of distribution are frequently found in sampling and in surveys in many areas, for instance in examination results and in quality control.

The standard deviation shows the dispersion of values around the arithmetic mean. The greater the dispersion the larger the standard deviation. Therefore a distribution with an arithmetic mean of £10 and a standard deviation of £4 has a greater or wider dispersion than a similar distribution with a standard deviation of £2. This can be shown graphically by saying that Curve *A* (in Figure 10.9) has a wider spread or dispersion than Curve *B*. They both have the same number of items and share the same arithmetic mean:

In Figure 10.9 both curves are of the 'normal curve' type. The further away a distribution is from this type the more difficult it becomes to interpret the standard deviation accurately.

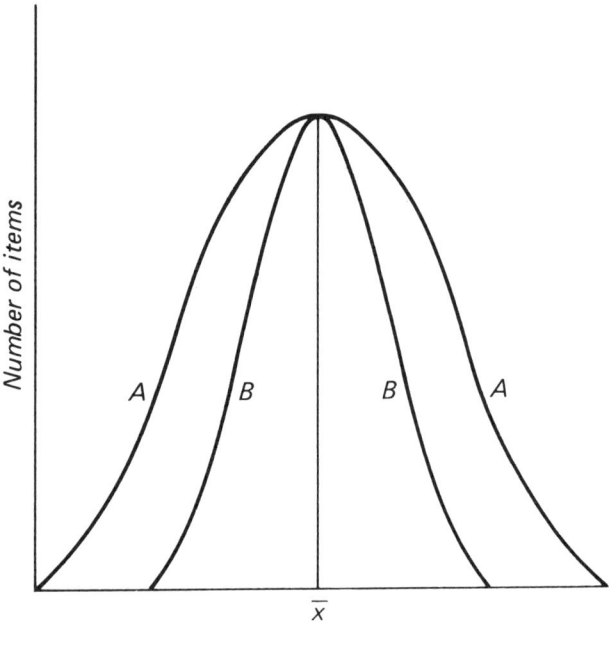

Fig 10.9 *distributions*

Fig 10.10 *the normal curve*

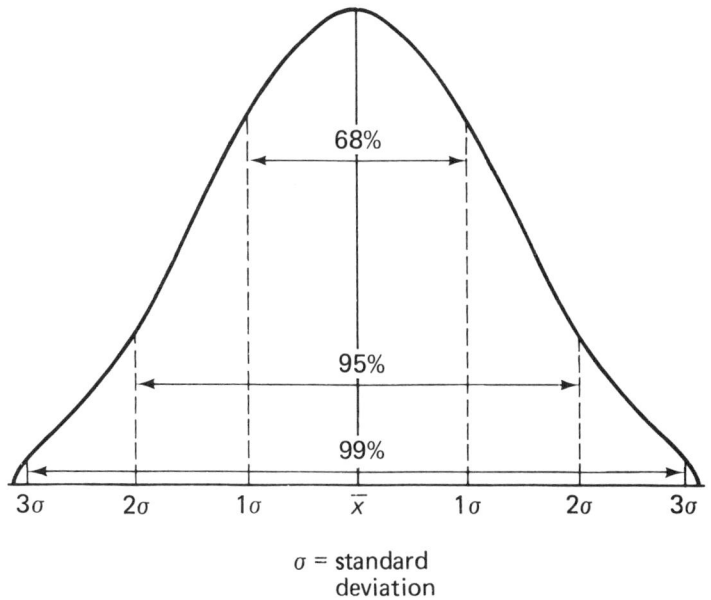

σ = standard deviation

With a normal curve it has been calculated that it is possible to mark off the area under the curve into certain proportions. This can be seen in Figure 10.10:

In this curve the arithmetic mean is shown in the centre with three standard deviations marked off on either side. It has been calculated that approximately 68% of the items of a distribution will lie within one standard deviation on either side of the arithmetic mean (two standard deviations in all). Within two standard deviations (or 1.96 standard deviations) on either side of the mean will lie approximately 95% of the items; and within three standard deviations on either side of the mean will lie approximately 99% (in fact 99.74% or nearly all the items) of the items in the distribution.

These are approximations, but this means that it is possible to arrive at a fairly clear picture of a distribution if the arithmetic mean, the standard deviation and the number of items are known.

For example: with a mean of £20, a standard deviation of £5 and 500 items in the distribution, it is possible to draw the curve in Figure 10.11. This curve is constructed by putting in the mean at the centre of the horizontal axis (£20), drawing in the vertical to represent the number of items (500) and then marking the horizontal scale by adding and subtracting the standard deviation (£5) from the mean (£20 ± 5 for one standard deviation,

Fig 10.11 *proportions of the normal curve*

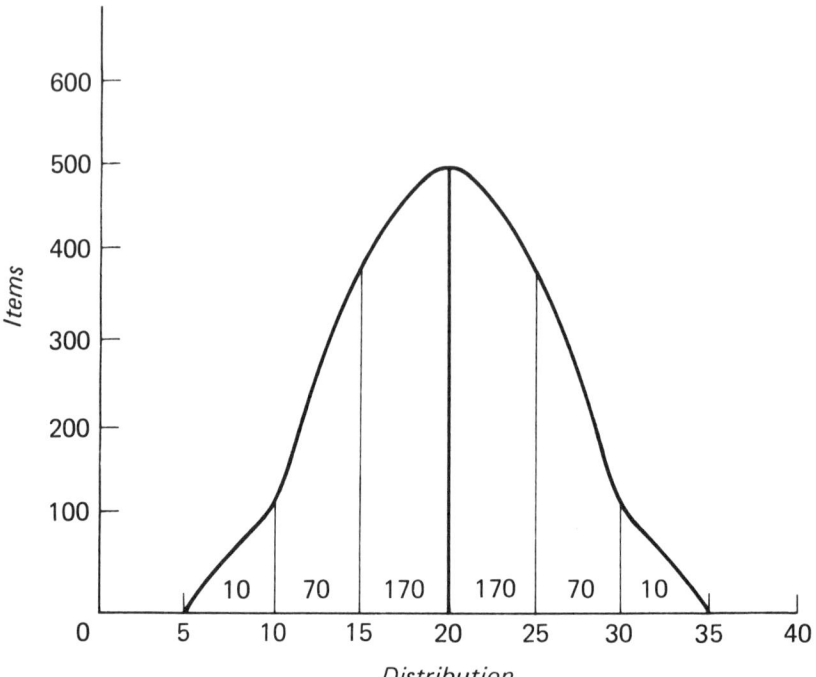

£20 ± 10 for two standard deviations and £20 ± 15 for three standard deviations). The shape of the curve can then be arrived at approximately by assuming that 68% of the items lie within one standard deviation on either side of the mean (say 340 out of 500 items); 95% of the items lie within two standard deviations of the mean (a total of approximately 480 out of 500 items); and 100% of the items lie within three standard deviations of the man. Strictly, the number of items under each section of the curve should be proportional to the area of that section.

(b) The calculation of the standard deviation

To calculate the standard deviation, the method used is to take the deviation of each value from an average and then calculate an average from these deviations in order to control for the number of cases involved. The problem is that if the actual deviations from the arithmetic mean are added together the result will always be zero, because the positive and negative deviations will cancel each other out.

For example: for the values 5, 6, 8, 11, 15, the arithmetic mean is 9 (45 ÷ 5). The deviation of each value from 9 is shown in Table 10.3.

Table 10.3 **deviations**

Value	Deviation	
5	− 4	
6	− 3	
8	− 1	
11		+ 2
15		+ 6
	− 8	+ 8

It can be seen in Table 10.3 that the positive and negative deviations cancel each other out. Therefore in order to obtain a measure of dispersion around the mean the negative signs must be removed. There are two methods of doing this:

(i) By ignoring the signs and taking the absolute values of the deviations (16 in Table 10.3). This method is used to arrive at the *mean deviation*. The mean deviation is the arithmetic mean of the absolute differences of each value from the mean (16 ÷ 5 = 3.2). The mean deviation is seldom used, because it is not mathematically very useful compared with the standard deviation.
(ii) By squaring the deviation. This has the effect of making the negative deviations positive. The square root of the result is taken to cancel out the squaring of the deviations. This is the method used to calculate the standard deviation.

Therefore the standard deviation is calculated by adding the square of the deviations of the individual values from the mean of the distribution, dividing this sum by the number of items in the distribution and finding the square root of the result. Taken step by step this is less complicated than it sounds:

(i) The deviation of one figure from another means the difference between them. This method has been used for calculating the arithmetic mean (Section 9.3).

For example: a survey of the weekly expenditure on transport for five families gives the results shown in Table 10.4.

Table 10.4 **transport expenditure**

Families	Weekly expenditure on transport (£)	Deviation from the assumed mean (25)	
1	15	− 10	
2	25	0	
3	26		+ 1
4	20	− 5	
5	14	− 11	
		− 26	+ 1
		= − 25	

In Table 10.4 the assumed mean is £25 above the real mean over a total of five values.
Therefore

$$£\frac{-25}{5} = -5$$

$$£25 - 5 = \underline{£20}$$

Therefore the arithmetic mean is £20. The first step in calculating the standard deviation is to calculate the arithmetic mean.

(ii) The second step is to square the deviation of the value of each item from the arithmetic mean (Table 10.5).

In Table 10.5 the sum of the squared deviations (the sum of the squares) is £122.

(iii) This sum (£122) is divided by the number of items in the distribution:

$$£\frac{122}{5} = £24.4$$

(iv) The final step is to find the square root of this figure:

$$\sqrt{24.4} = \underline{£4.9} \text{ approximately}$$

Therefore the standard deviation of this distribution is £4.9 and the arithmetic mean £20.

Table 10.5 **squared deviations**

Families	Weekly expenditure on transport (£)	Deviation from arithmetic mean (20)	Deviation squared
1	15	− 5	25
2	25	+ 5	25
3	26	+ 6	36
4	20	0	0
5	14	− 6	36
			122

The basic formula for calculating the standard deviation is:

$$\sigma = \sqrt{\frac{\Sigma x^2}{n}}$$

where

σ represents the standard deviation (sometimes S or S.D. is used)
x^2 represents the sum of the squared deviations from the mean
n is the number of items (f in a frequency distributions).

The distribution in Table 10.5 is unimodal and fairly symmetrical. By drawing the histogram and polygon it is clear that it is partly skewed (Figure 10.12). The fact that this distribution is partly skewed is reflected by the distribution of items. While 3.4 of the items represent 68% of the total number of items (i.e. families), they cover approximately 75% of the values (the three central values: 25, 26 and 20 and say 4 from the other two values — making 75 out of 100). Also, 4.8 of the items, representing 95% of the total number of items in the distribution, cover approximately 98% of the values.

Looking at the standard deviation in this way makes it clear whether the calculation is reasonable or not (see Section 10.4(a)). To put it another way, one standard deviation on either side of the mean should cover about 68% of the items in the distribution as has been seen above. In fact in this example the range £15 to £25 (£20 plus and minus £5) includes the values of three out of five items. This is 60%, which is very close to 68%, given the small number of items involved and the fact that the distribution is skewed to some extent. This indicates that the standard deviation calculated as £4.9 for this distribution is reasonable.

If the standard deviation in the example had been calculated as £2, the

Fig 10.12 *histogram of transport survey*

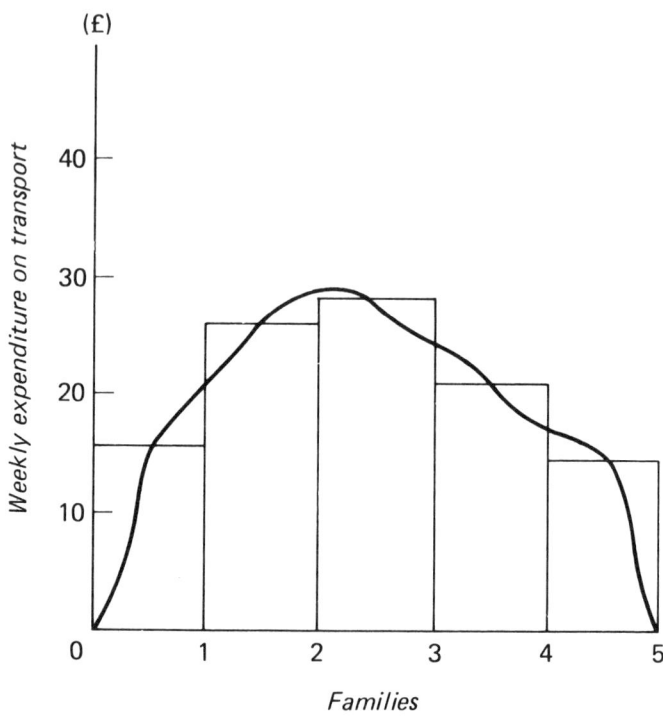

limits around the arithmetic mean (£20) for one standard deviation would have been £18 to £22. This would have included the value of only one item (family number 4 with a weekly expenditure on transport of £20), and therefore the standard deviation would appear to be too small.

On the other hand, if the standard deviation had been calculated as £8, the limits around the arithmetic mean for one standard deviation would have been in the range £12 to £28. This range would include all the values in the distribution, and therefore the standard deviation would appear too large.

This summarisation of a distribution is particularly important where there are a large number of items and where there is a grouped frequency distribution. Computer programmes often produce either the standard deviation as a summary figure, or a figure which is based on the standard deviation (such as the Variance and the Coefficient of Variability in Sections 10.5 and 10.6).

(c) Calculation of the standard deviation of a grouped frequency distribution
This is more complicated than the example used so far, but the overall method is similar and the interpretation is the same.

The method of calculation used here is similar to that used to find the arithmetic mean for a group distribution (see Section 9.4). It would be possible to use deviations from the mid-points of the classes, but to avoid using unwieldy figures in the last two columns it is easier to use units of class intervals (particularly with even class intervals). These 'class-interval' units have to be reconverted to the units of measurement at the end.

The formula for calculating the standard deviation of a grouped frequency distribution is:

$$\sigma = \sqrt{\frac{\Sigma f dx^2}{\Sigma f} - \left(\frac{\Sigma f dx}{\Sigma f}\right)^2} \times \text{class interval}$$

where f = the frequency
  d_x = the deviation from the assumed mean

$\sqrt{(\Sigma f dx^2/\Sigma f)}$ in fact represents the assumed deviation. $(\Sigma f dx/\Sigma f)^2$ is the correction necessary to arrive at the real standard deviation. It is always subtracted from the assumed standard deviation.

Therefore $\Sigma f dx^2$ is the sum of the frequencies + deviations from the assumed mean squared. It is found, in Table 10.6, by multiplying columns 4 and 5, but it could be found by squaring column 4 ($-4^2 = 16$) and multiplying the result by column 2 ($16 \times 2 = 32$). It is important to note that $\Sigma f dx^2$ is the frequencies × deviations squared.

Table 10.6 shows the simplest method of calculation and makes it possible to calculate the arithmetic mean and the standard deviation from the same table. It gives the results of a survey into the expenditure on transport each month by 100 families.

In Table 10.6 (p. 168) the calculations are based on the assumed mean (£45). The arithmetic mean and standard deviation are:

$$\bar{x} = 45 + \frac{21}{100} \times 10 = 45 + 2.1 = \underline{£47.1}$$

$$\begin{aligned}
\sigma &= \sqrt{\frac{\Sigma f dx^2}{\Sigma f} - \left(\frac{\Sigma f dx}{\Sigma f}\right)^2} &&\times \text{class interval} \\
&= \sqrt{\frac{341}{100} - \left(\frac{21}{100}\right)^2} &&\times 10 \\
&= \sqrt{341 - 0.0441} &&\times 10 \\
&= \sqrt{3.3659} &&\times 10 \\
&= 1.835 \times 10 &&= \underline{£18.35}
\end{aligned}$$

Table 10.6 standard deviation of a grouped frequency distribution

(1)	(2)	(3)	(4)	(5)
Expenditure (in classes of 10) (£)	Number of households (f)	Deviation of classes from class of assumed mean (d_x)	Frequency × deviations from assumed mean (2) × (3) (fd_x)	Frequency × deviation from assumed mean, squared (3) × (4) (fd_x^2)
0 and under 10	2	−4	−8	32
10 and under 20	6	−3	−18	54
20 and under 30	12	−2	−24	48
30 and under 40	15	−1	−15	15
40 and under 50	18	0	0	0
50 and under 60	20	+1	+20	20
60 and under 70	17	+2	+34	68
70 and under 80	8	+3	+24	72
80 and under 90	2	+4	+8	32
	100		−65 +86	341
	$\Sigma f = 100$		= +21 $\Sigma fd_x = +21$	$\Sigma fd_x^2 = 341$

The arithmetic mean of £47.10 and the standard deviation of £18.35 can be used to compare the dispersion in this data and other data, providing the distributions are of the same kind and are measured in the same units.

Notice that one standard deviation on either side of the mean (47 − 18 = 29, 47 + 18 = 65) gives limits of about £29 and £65. These limits include the frequencies 15, 18, 20 and half of 17 (say 8). This is a total of 61 out of 100, which is close to 68%.

Two standard deviations on either side of the arithmetic mean gives limits of about £11 and £83. These limits include all the frequencies except the first and last and therefore about 96% of the distribution. This is very close to 95%.

In fact the distribution is not perfectly symmetrical and this accounts for the difference between 61% and 68%, and 96% and 95%. However, it is clear that the result arrived at for the standard deviation of this distribution is reasonable.

10.5 THE VARIANCE

The variance is the average of the square of the deviations.

For example: Table 10.7.

Table 10.7 **the variance**

Values	Deviation from arithmetic mean (11)	Deviation squared
5	− 6	36
6	− 5	25
10	− 1	1
14	+ 3	9
20	+ 9	81
		152

$$\text{Variance} = \frac{152}{5} = \underline{30.4}$$

Thefore the variance is calculated in the same way as the standard deviation, but without the final step of finding the square root. The standard deviation for the distribution in Table 10.7 would be approximately 5.5. So the variance can be interpreted as the square of the standard deviation (σ^2).

10.6 THE COEFFICIENT OF VARIATION

It is sometimes desirable to compare several groups with respect to their relative homogeneity in instances where the groups have very different means. In these circumstances it might be misleading to compare the absolute magnitudes of the standard deviations. It might be more relevant and of greater interest to look at the size of the standard deviation relative to that mean. A measure of relative variability can be obtained by dividing the standard deviation by the mean.

Therefore the coefficient of variation (V) is the standard deviation divided by the mean:

$$V = \frac{\sigma}{\bar{x}}$$

For example: if two groups of data are compared the results could be as follows:

Group A $\bar{x} = 15$ $\sigma = 3$ $V = \dfrac{3}{15} = 0.2$

Group B $\bar{x} = 35$ $\sigma = 5$ $V = \dfrac{5}{35} = 0.14$

The coefficient can be made into a percentage by multiplying by 100. The coefficient of variation for Group A then becomes 20% and for Group B becomes 14%.

10.7 CONCLUSIONS

A measure of dispersion helps to describe a distribution better than an average on its own. It can show whether or not the figures in the distribution are clustered closely together or well spread out.

Dispersion may be as important as an average because in many areas it is changes in the spread of a distribution which are of interest as much as changes in the average. For instance, economists may be as interested in changes in the distribution of incomes as in changes in the average income.

The standard deviation is the most important of the measures of dispersion due to its mathematical properties (especially in sampling theory).

ASSIGNMENTS

1 Obtain a price list for any consumer goods where there are at least ten different prices. Calculate the arithmetic mean, the median, the modal price, the price range and the interquartile range.

Discuss how far the calculations have helped to summarise the list of prices. Comment on which calculation might be the most useful one to make if you were to consider buying the consumer goods.

2 The following table summarises the performance of two companies on wages and hours. From this information write a profile of the two companies. On the basis of this evidence, comment on which of these two companies might be the better one by which to be employed.

	Company A		Company B	
	Arithmetic mean	Standard deviation	Arithmetic mean	Standard deviation
Weekly wages	80	5	90	15
Weekly overtime earnings	20	10	20	15
Hours per week	35	1	40	3
Number of days holiday a year	20	4	15	6

3 A company has the following distribution of overtime earnings. Calculate the arithmetic mean and the standard deviation. Discuss what the value of the standard deviation shows about the weekly overtime earnings of this company.

Weekly overtime earnings £	Number of staff
5 and under 10	3
10 and under 15	10
15 and under 20	25
20 and under 25	8
25 and under 30	4

4 The following data shows the monthly expenditure on advertising of the branches of an electrical goods company during 1979:

Monthly advertising expenditure (£)	Number of branches
400 and less than 600	4
600 and less than 800	52
800 and less than 1000	110
1000 and less than 1200	70
1200 and less than 1400	64
1400 and less than 1600	47
1600 and less than 1800	43
1800 and less than 2000	10

From this data calculate the median monthly expenditure and explain what it indicates about the branches advertising expenditure in 1979. Also, calculate the interquartile range and explain the purpose of this calculation.

5 The following table shows the results of information collected by a company selling consumer products directly to retailers:

Average, number of orders taken for month by individual salesman	Number of salesmen
10 and under 20	2
20 and under 30	4
30 and under 40	9
40 and under 50	11
50 and under 60	12
60 and under 70	35
70 and under 80	30
80 and under 90	16
90 and under 100	1

Calculate the range, the standard deviation and the coefficient of variation.

Discuss how far these measures of dispersion help to interpret the information the company has collected.

6 Discuss the purpose of:

(i) calculate a measure of dispersion;
(ii) identifying differently shaped distributions.

CHAPTER 11

STATISTICAL DECISIONS

11.1 ESTIMATION

Much of the coverage of the chapters up to this stage has been concerned with the collection of statistical information and maximising the comprehension of the groups of figures collected. The next step is to increase comprehension of the populations from which the figures have been collected.

This is the difference between observing vehicles using a section of road and describing what happens, and using this group as a sample to describe all vehicles including all those not observed.

Statistical estimation is concerned with finding a statistical measure of a population from the corresponding statistic of the sample; for instance, whether the arithmetic mean of a sample is a good estimate of the arithmetic mean of the population. It is an 'estimate' because it is never certain that a sample will be an exact minature copy of the population itself.

As was seen in Chapter 5, the underlying objective of sampling is to describe the population from which the sample is taken. It was established there that sampling is based on a random selection of items and on the theory of probability. In this chapter it is possible only to provide a very brief discussion of probability and of statistical estimation to give some idea of the basis of statistical induction; that is the process of drawing general conclusions from a study of representative cases.

11.2 PROBABILITY

The origins of the concept of probability lie in a simple mathematical theory of games of chance and can be traced back to the 1650s when a French gambler consulted a well-known mathematician, Pascal, presumably to find out how often he might win.

Uncertainty is common to games and to business and probability can provide analytical tools for measuring and controlling aspects of uncertainty. The use of probability in decision making can be based on an analysis of the past behaviour of sales or production and probabilities assigned to possible outcomes.

This is not the same as the subjective approach used by bookmakers when fixing the odds for a particular horse to win a race. In this approach, initially the odds will be determined by the personal view of the bookmaker, and will be modified as the time for the race approaches according to the subjective views of the punters.

A definition of probability is that the probability of an event is the proportion of times the event happens out a large number of trials.

For example:

(i) The probability of a date chosen at random from a calendar being a Monday would be 1 in 7, or 1/7 or 0.143 or 14.3%.
(ii) The probability of tossing a head when throwing a coin would be 1 in 2 or $\frac{1}{2}$ or 0.5, or 50%.

In (i), if a very large number of dates were looked at, then each day of the week would turn up the same number of times. Therefore Monday would turn up on 1/7 of the occasions and, by definition, this is the probability.

In everyday conversation words like 'chance', 'likelihood' and 'probability' are used to convey information, but they are not related to the above definition of probability. They are subjective in the same way as the bookmaker's approach, they are estimates based on experience and knowledge of the factors surrounding the situation.

It can be argued that most decisions are made on a qualitative rather than a quantitative basis, even in business. Actions are taken because certain events are very likely to happen, not because their probability is precisely say 97.3%.

As was suggested in Chapter 1, statistics are introduced to support decisions, to provide evidence and to narrow the area of disagreement.

Therefore probabilities can be regarded as relative frequencies; the proportion of time an event takes place. The relative frequency is the probability that the particular event will happen.

For example: when it is said in a manufacturing company that the 'probability of an order being complete on time is 0.8', it is meant that on the basis of past experience 80% of all similar orders were met on time.

The probability of obtaining heads when a coin is tossed is 50%. This does not mean that when a coin is tossed 10 times, 5 heads are always obtained. However, if the experiment (or sample or event) is repeated a large number of times then it is likely that 50% heads will be obtained.

The greater the number of throws the nearer the approximation is likely to be. In this sense probability is a substitute for certainty.

A more advanced study of probability attempts to provide mathematical support for assigning probabilities to events.

11.3 THE NORMAL CURVE

The normal curve is a bell-shaped symmetrical distribution (see Section 10.1 and Section 10.4) and its shape will depend on the mean and the standard deviation. A normal curve may be very tall, or very flat, as shown in Figure 11.1. On both of the curves in Figure 11.1 the arithmetic mean is

Fig 11.1 *normal distributions*

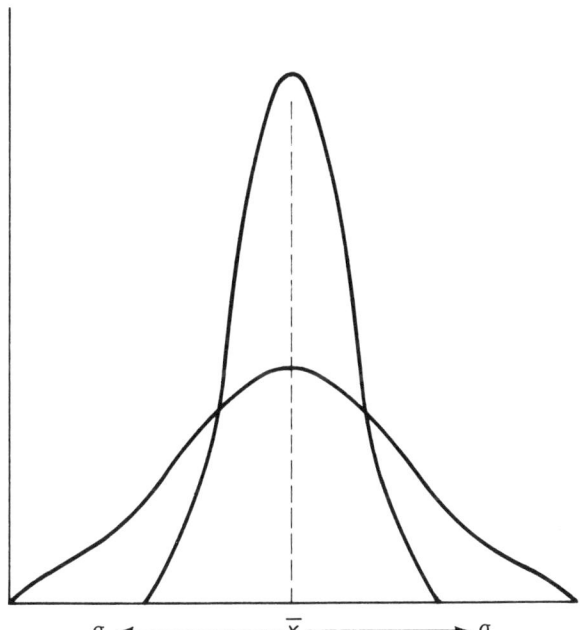

shown in the centre. In both cases approximately 68% or two-thirds of the distribution will lie within one standard deviation on either side of the arithmetic mean; approximately 95% or 19 out of 20 of the items will lie within 2 standard deviations on either side of the mean; almost all the items will lie within 3 standard deviations of the mean (see Section 10.4). Sometimes 2.58 standard deviations on either side of the mean are used to include 99% of the items, while 99.74% of the items are represented by 3 standard deviations on either side of the mean.

For example: Figure 11.2.

Fig 11.2 *area under the curve*

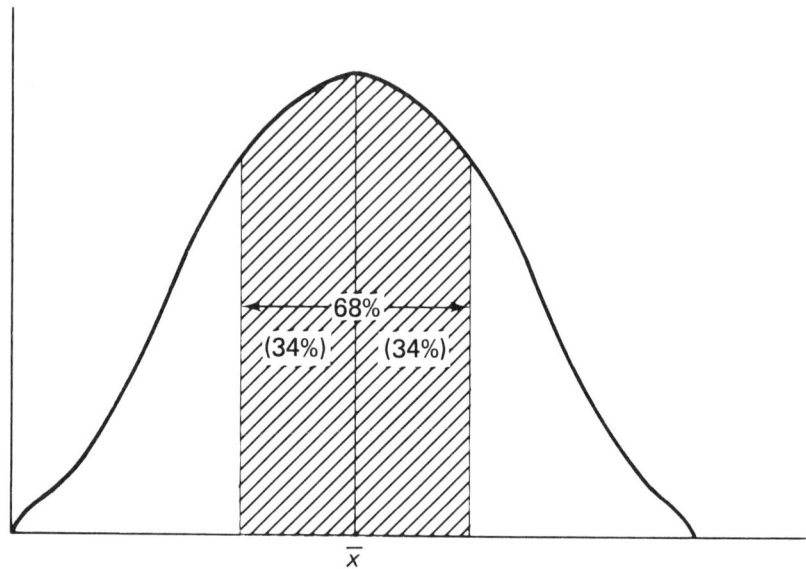

Mathematicians have computed the area that lies under any part of the curve (a table of these areas is given in Appendix A.2). In the area table, the area between the mean line, the curve, the horizontal axis and a vertical line measured from the mean in units of standard deviations is given as a decimal of the total area enclosed by the curve.

For example: if the line lies $1\frac{1}{2}$ σ from the mean, the table shows that the area enclose is 0.4332 or 43.32% of the total area. Or 86.64% for $1\frac{1}{2}$ σ on either side of the mean.

The curve which represents these areas is referred to as a 'standard normal distribution'. To apply a standard normal distribution to an actual distribution, the actual distribution is superimposed on the standard normal distribution scale so that the actual distribution mean coincides with the 'zero' (or the mean) and the actual distribution values coincide with the appropriate σ points.

For example: a distribution with a mean of 400 cm and a standard deviation of 20 cm will have 400 cm at 0 and the 400 − 20 and 400 + 20 values lying at the −1 σ and +1 σ points (Figure 11.3).

Fig 11.3 *a normal distribution*

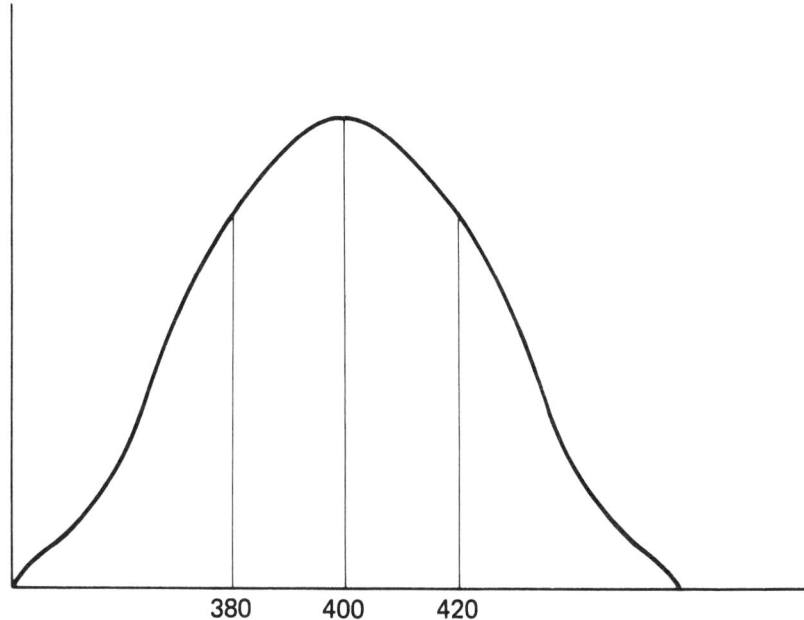

The standard normal distribution is a way of stating an actual value in terms of its standard deviation from the mean in units of its standard deviation (illustrated in Figure 11.3). This new value can be called a Z value (Figure 11.4). Z is the distance any particular point lies from the mean, measured in units of standard deviation:

$$Z = \frac{\text{the value} - \text{the mean}}{\text{standard deviation}} = \frac{x - \bar{x}}{\sigma}$$

where x = any particular value.

In the example in Figure 11.3, the Z value of 360 cm would be:

$$Z = \frac{360 - 400}{20} = \frac{-40}{20} = -2$$

Therefore the 360 cm value lies two standard deviations below the mean of the distribution, and this will include approximately 47.5% (from the area tables) of the distribution.

Therefore if the mean and the standard deviation are known, and what is required is an idea of the percentage of the distribution lying between the mean and a particular value, the Z values can be used.

For example: if a distribution has a mean of 20 and a standard deviation of 5, the area under the curve between the mean and 28 would be:

$$Z = \frac{28 - 20}{5} = 1.6$$

The area table shows that when $Z = 1.6$ the area is 0.4452. Therefore the area lying under the curve is 44.5%. Therefore 44.5% of the distribution will lie between 20 and 28.

Fig 11.4 *the Z value*

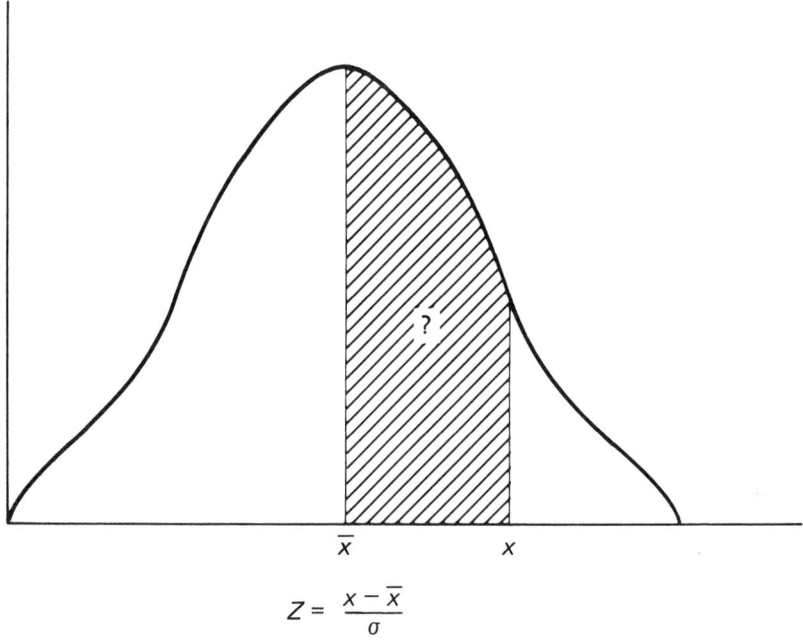

$$Z = \frac{x - \bar{x}}{\sigma}$$

11.4 PROBABILITY AND THE NORMAL CURVE

The fact that 68% of the total area under a normal curve lies within one standard deviation either side of the mean makes it possible to say that 2 items out of 3 lie within one standard deviation of the mean. Therefore if one item was selected at random from a distribution, there would be a 2 in 3 chance of it being one of the items lying within one standard deviation of the mean. This would be correct 2 times out of 3 or it would be likely with a '68% level of confidence'.

Similarly, since 95% of the area lies within two standard deviations of the mean, one item selected at random would lie within these limits 19 times out of 20. It would be possible to say this at a '95% level of confidence'. Nearly all the items would be within three standard deviations of the mean with a '99% level of confidence'.

Therefore, if a sample is selected at random from a population, there is a good chance or probability that it will represent the population from which it is drawn. Describing a population from observing a small sample is not a matter of guess work providing:

(i) the sample is taken at random,
(ii) it is understood that there is always some degree of error in sampling.

A sample will not be an exact replica of the population, because chance plays an important part in random selection. It is likely that the sample mean and standard deviation will differ a little from the population mean and deviation. The bigger the sample the less chance there is of selecting a group of items that is unrepresentative of the population.

In general it can be stated that:

(i) the best estimate of the population mean is the sample mean;
(ii) the best estimate of the population standard deviation is the sample standard deviation.

11.5 STANDARD ERROR

The fact that there is always some degree of error in sampling means that it is useful to have a measure which provides an indication of the extent to which sample means deviate from population means.

If a large number of samples are taken from a population, which is normally distributed, most of the sample means will be the same or very similar. Occasionally a sample will by chance contain an undue number of high or low values, so that its mean will be above or below other means. If all these means are graphed they will look like a normal curve, with most of the means in the centre but some on either side. This 'sampling distribution of the mean' will be normally distributed and the average of these means will be equal to (or 'the best estimate of') the true population mean.

The standard deviation of this distribution is called the standard error. It is found by the formula:

$$\text{Standard Error (S.E.)} = \frac{\text{standard deviation of the sample}}{\sqrt{\text{sample size}}}$$

$$= \frac{\sigma}{\sqrt{n}}$$

where n = sample size.
(In this formula σ may be replaced by S because this is the estimate of the population standard deviation.)

Therefore:

(i) since 95% of the items in a normal distribution lie within 2 standard deviations of the mean of the distribution,
(ii) since the distribution of the means is normal with a mean equal to the population mean,
(iii) then 95% of the means of all samples must lie within two standard errors of the true mean of the population,
(iv) and therefore if a single sample is taken, 19 times out of 20 the sample mean will lie within two standard errors of the true mean of the population. Or in other words, 19 times out of 20 the true mean of the population cannot lie more than two standard errors from the mean of the sample.

Notice that the accuracy of the estimate is independent of population size. A population of 2 million does not require a bigger sample than a population of 20,000. Accuracy depends on the sample size and the variability of the characteristics measures (see Section 5.4).

For example: a sample of 100 with a standard deviation of 5 will have a standard error of:

$$\text{S.E.} = \frac{\sigma}{\sqrt{n}} = \frac{5}{\sqrt{100}} = \frac{5}{10} = 0.5$$

The accuracy of the sample can be increased by expanding the size of the sample. A sample of 200 will have a standard error of:

$$\text{S.E.} = \frac{5}{\sqrt{200}} = \frac{5}{14.2} = 0.35$$

To halve the standard error in the original sample (0.5), the sample size has to be increased by four times:

$$\text{S.E.} = \frac{5}{\sqrt{400}} = \frac{5}{20} = 0.25$$

The variability of any given population may change over time. When the electorate of the UK was felt to be predictable in its voting patterns (in the 1950s and 1960s) a sample of 2000 electors was throught to provide a reasonably accurate prediction of election results. In more recent years it has been felt that the variability and volatility in the electorate has increased and therefore larger samples are required to provide a reasonable prediction of election results. Notice that this is not because of an increase in the size of the population, but because a larger sample is required to represent this population.

The theory behind standard error and sampling distributions are part of a mathematical theory called the central limit theorem (see Section 5.4). Detailed discussion of this theorem lies outside the scope of this book. However, as a result of this theorem it can be accepted that the sampling distribution is normal even if the population frequency distribution is not normal, providing that the sample size is sufficiently large (greater than 30). This means that it is possible to take a sample from any population and apply methods of estimation.

11.6 TESTS OF SIGNIFICANCE

In statistics it may be that some fact or theory is believed to be true, but when a random sample is taken the results do not wholly support the fact or theory.

The difference between the belief or hypothesis and the sample result may be because:

(i) the original theory or hypothesis was wrong;
(ii) the sample was one-sided.

Significance tests are aimed at revealing whether or not the difference could be reasonably ascribed to chance factors operating at the time the sample was selected. If the difference cannot be explained as being due solely to chance the difference between the theory and the sample result is said to be statistically significant.

To test a hypothesis it is possible:

(i) to collect information on the whole population and then accept or reject the hypothesis with complete certainty;
(ii) to take a random sample, if the population is too large for a full survey, and test the null hypothesis.

The 'null hypothesis' is the assumption that there is no difference between the hypothesis and the sample result.

For example: if the theory is that a group of employees receive an average wage of £130 a week, then the null hypothesis (Ho) is that the population mean (μ: pronounced 'mew') is £130. The opposite theory (Hi) is that it is not £130:

Ho: μ = £130 (the population mean is £130)
Hi: $\mu \neq$ £130 (the population mean is not £130)

A random sample is taken and the results show:

n = 100 $\bar{x}$ = £123 σ = £30

If the population mean is £130, then 95% of the means of all samples will fall within two standard errors of this figure (as was seen in the sampling distributions of the means). If the sample mean, £123, is not within two standard errors of £130, then the population mean is probably not £130 unless the sample mean is the one in twenty that is one-sided.

The standard error of the sample results is:

$$S.E. = \frac{\sigma}{\sqrt{n}} = \frac{30}{\sqrt{100}} = \frac{30}{10} = 3$$

The critical values are the limits in which the sample mean will fall 95% of the time if the population mean is correct.

The values are found by taking the population mean, plus and minus two standard errors:

The critical values = £130 ± 2 × 3
= £130 ± 6
= £124 to £136

The sample mean = £123. This does not lie within the critical values; therefore Ho (the null hypothesis) is rejected at the 5% level of significance. The difference between the assumed population mean (£130) and the sample mean (£123) can be said to be significant.

From this test it can be stated that there is evidence to suggest that the population mean is not £130 a week.

If the sample mean had fallen within the critical values the Ho would be accepted at the 5% level and the difference between μ and $\bar{x}$ would not be significant. This would not mean that the hypothesis was 'proved'. It would mean that there was evidence to suggest that the population mean could be £130 a week.

It is possible to expand the critical values by taking the 1% level of significance where it is suggested that between 99% and 100% of the items (sample means) will lie within three standard errors of the population mean.

$$\text{Critical values} = £130 \pm 3 \times 3$$
$$= £130 \pm 9$$
$$= £121 \text{ to } £139$$

At this level of confidence the null hypothesis (Ho) is accepted. Therefore there is evidence to suggest that the population mean could be £130 a week, because the sample mean (£123) falls within the critical values.

However, by widening the critical values there is an increased probability that the null hypothesis is being accepted when it is in fact wrong (that is, an increased chance of making a Type II error).

There are two types of error which can be made in testing significance which are called Type I and Type II:

(i) a Type I error is rejecting the null hypothesis (Ho) when it is in fact true,
(ii) a Type II error is accepting the null hypothesis when it is in fact false.

The level of significance can be said to equal the probability of making a Type I error. If the critical values are expanded there is an increased probability of making a Type II error.

The level of significance chosen and the interpretation of the results of the test are a matter of judgement. If more evidence is required further random samples can be taken.

11.7 CONFIDENCE LIMITS

An alternative test to the significance test is instead of testing specific values of the population mean, to construct an interval that will, at specified levels of probability, include the population mean.

For example: a random sample taken from the weekly wages of a group of employees provides the following results:

$$n = 100 \quad \bar{x} = £20 \quad \sigma = £20$$

These results can be used to set 95% confidence limits on the unknown population mean μ.

$$\text{The standard error of these results} = \frac{20}{\sqrt{100}} = \frac{20}{10} = 2.$$

$$95\% \text{ confidence limits} = \bar{x} \pm 2 \times \text{S.E.}$$
$$= £20 \pm 2 \times 2$$
$$= £20 \pm 4$$
$$= £16\text{-}£24$$

It can be said (approximately) that there is a 95% chance that this range will include the unknown population mean (μ).

Narrower limits can be achieved by increasing the size of the sample. If the sample size was 400 and the results were the same, then:

$$\text{S.E.} = \frac{20}{\sqrt{400}} = \frac{20}{20} = 1$$

$$95\% \text{ confidence limits} = £20 \pm 2 \times 1$$
$$= £18\text{-}£22$$

Setting 99% confidence limits increases the likelihood of including the population mean in the range, but it also increases the size of the interval.

With $n = 100$:

$$99\% \text{ confidence limits} = £20 \pm 3 \times 2$$
$$= £14\text{-}£26$$

With $n = 400$:

$$99\% \text{ confidence limits} = £20 \pm 3 \times 1$$
$$= £17\text{-}£23.$$

11.8 STATISTICAL QUALITY CONTROL

One application of sampling methods and of probability is statistical quality control. This involves sampling the output of a manufacturing process to ensure that the quality of the commodities being produced conforms to specified standards.

Mass production techniques involve making sure that each item of output has standard dimensions within certain limits of an ideal standard. Products can be checked when they come from the machine, but by that time the damage may have been done and a whole batch may be lost. It is more economical to take a series of samples during production in order to try to discover any fault at the earliest possible moment.

With any commodity there may be variations in a number of dimensions of the commodity, such as the mean length. Frequent samples, usually taken automatically, will check on this. If a frequency distribution of the mean length of products is taken, the curve is likely to be bell-shaped, or similar to the normal curve. Most components will be the right size but a few will be too large or too small.

For example: if the ideal length of a component is 5 cm and there is a tolerance of 0.004 cm, then components of 4.996 cm and of 5.004 cm will be acceptable.

The frequency curve of a sample may appear as in Figure 11.5:

Fig 11.5 *quality control curve*

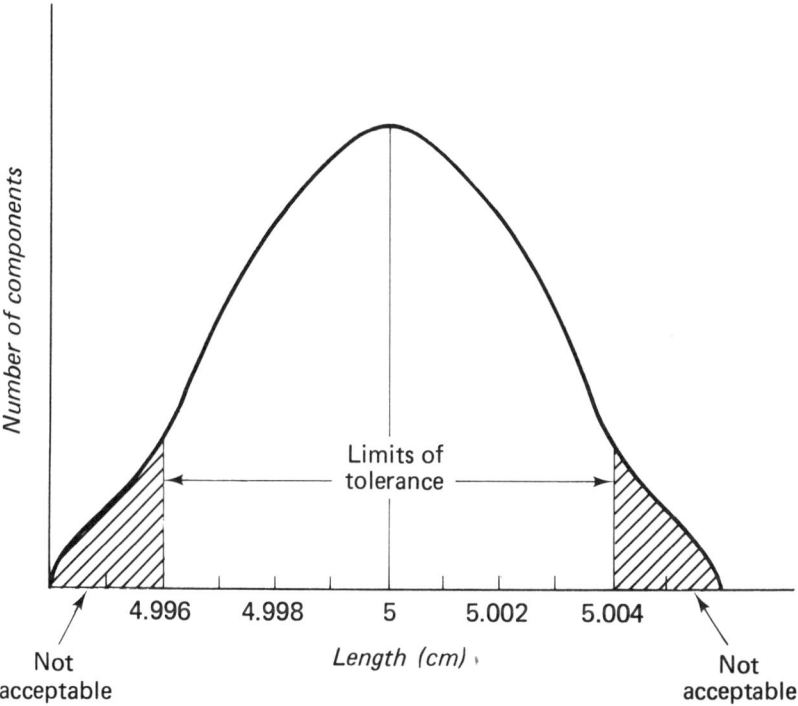

The samples taken are plotted on quality control charts to provide a visual indication of quality variations (Figure 11.6). In Figure 11.6 the action or control limits are limits of standard error and therefore are confidence limits. One important advantage of the control chart is that output from a production process often deteriorates progressively and this is made apparent in a chart by the trend of the plotted points. In the example in Figure 11.6 the points are moving out towards the upper warning limit and therefore the production line should be investigated.

More detailed and complicated techniques can be applied to the control of quality and important aspects of quality control in practice involves engineering problems. It is introduced in this chapter as an example of sampling.

Fig 11.6 *quality control chart*

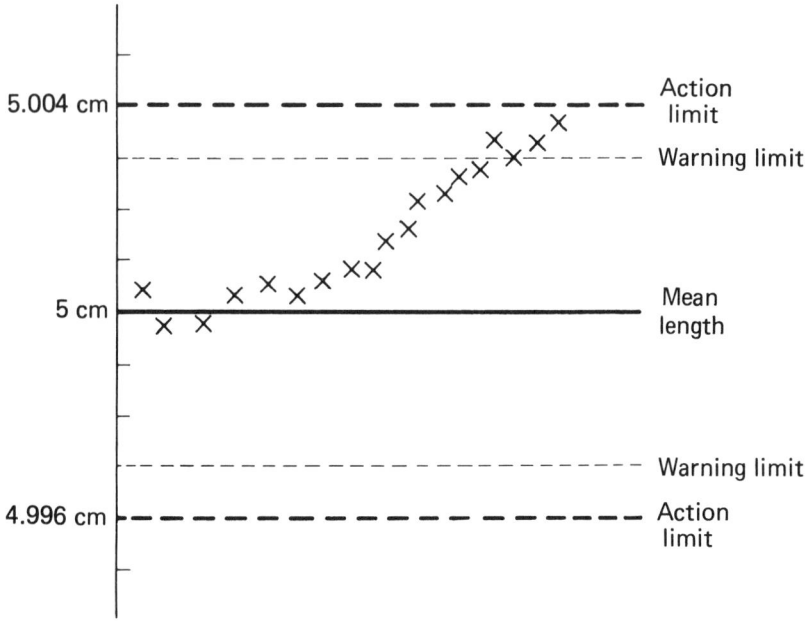

11.9 CONCLUSION

It can be argued that decisions are made on emotional and irrational grounds rather than on facts. Perhaps this depends on the type of decisions being made. For instance, if couples looked at the facts (costs, sleepless nights, etc.) before they decided to have children, the population might decline rapidly.

The decision to buy a car or a house has to be based on some facts. For instance the amount of money available through saving and borrowing must be a limiting factor. The size of the family and other factors may also influence the choice of car or house. Once these factors have been taken into account, the choice of a particular car or house may depend on personal preferences or chance factors and therefore not on clearly rational grounds.

Justification for such decisions will tend to be expressed in rational terms although sometimes people will admit that they bought a car because they liked the colour or shape, or a house because they 'fell in love with it'.

Business decisions are perhaps similar to this. The facts provide the limits within which preferences and 'hunches' have to work; and the facts provide the justification for the decision when it is being explained to the bank manager, the 'boss' or the shareholders.

Statistical decisions are concerned with the reliance which can be placed on facts, the confidence with which figures can be used and the interpretation of facts and figures.

ASSIGNMENTS

1 Discuss the problems of decision making. How are decisions made in business? How far can statistics help in business decision-making?

2 What is the meaning of standard error? Why is the concept important in sampling?

3 It is thought that the cost of transport to and from work of a group of employees is £100 a year. A sample is taken and the results show that of the 100 employees surveyed, their average annual transport costs are £80 a year with a standard deviation of £10. Is there any evidence from the sample results that the employees' transport costs could be £100 a year?

4 Discuss the importance in statistics of:

(i) probability.
(ii) the normal curve

5 Comment on the use of significance tests and confidence limits. How far do they help in making decisions?

6 Find examples of the use of quality control. Consider the methods used in these cases.

CHAPTER 12
COMPARING STATISTICS: INDEX NUMBERS

12.1 IN GENERAL

An index number is a measure designed to show average changes in the price, quantity, or value of a group of items, over a period of time.

The aim is to provide a device to simplify comparison over time, by replacing complicated figures by simple ones calculated on a percentage basis. Most index numbers are weighted averages and they have similar advantages and disadvantages to averages. For this reason many types of index have been developed, some of which are more suitable than others for particular purposes.

There are three general types of indexes (or indices):

(i) price indexes: which measure changes in prices;
(ii) quantity indexes: which measure production and output changes;
(iii) value indexes: which measure changes in the value of various commodities and activities.

For example: the average price of houses can be compared between two years:

Year	Average house prices (£)	Index number (price relative to 1976)
1976	18,000	$\frac{18,000}{18,000} \times 100 = 100$
1980	30,000	$\frac{30,000}{18,000} \times 100 = 166$

In this index 1976 is the base year (the year on which the price changes are based) and is written: 1976 = 100. The index for 1980 (166) indicates that average house prices have risen by 66% between 1976 and 1980.

Where one item is involved in comparisons between different periods of time, the price or value index is found by:

$$\text{Price index} = \frac{p_1}{p_0} \times 100$$

Where p_0 is the price in the base year (1976) and p_1 in the year to be compared (1980). Where more than one item is involved the calculation of an index is more complicated. In calculating an index to show changes in the cost of living, for example:

(i) some prices will rise, others fall,
(ii) prices are for different units (weights and quantities of goods),
(iii) different products are of varying importance in the cost of living,
(iv) households spend their money in different ways.

For example: an extreme case would be to measure changes in the cost of living by changes in the prices of two commodities, say bread and matches. It has to be assumed that there is an 'average' loaf of bread of a particular weight and size and an 'average' box of matches containing a certain number of matches. One household (A) may consume large amounts of bread but never buys matches, while another household (b) consumes small amounts of bread and buys several boxes of matches a week (perhaps because some members of the family are pipe-smokers!).

If the price of bread has risen from 20p an average loaf in 1978 to 30p an average loaf in 1980, this is a percentage increase (1978 = 100) of 50%.

If the price of matches has increased from 5p an average box in 1978 to 6p an average box in 1980, this is a percentage increase (1978 = 100) of 20%.

Therefore the average percentage increase in the cost of living could be calculated as:

$$\text{Price rise} = \frac{50 + 20}{2}\% = 35\%$$

Or it could be said that the index had risen from 100 in 1978 to 135 in 1980. However, there are a number of problems with both these statements:

(i) They are based on assumptions such as only two commodities, with an 'average' size.
(ii) Household A does not buy matches, but only buys bread and therefore it has had an actual increase in the cost of living of 50%, and the 35% figure does not represent it at all.
(iii) No allowance has been made for the various amounts of bread and

matches that household *B* buys in a particular time period. Even if this household buys three boxes of matches to one loaf of bread, it is still spending more on bread.

This illustrates the problem of trying to provide an average figure to represent changes involving large numbers of people and households or companies and goods. Despite these problems a very large number of indexes are produced, for prices, wages, production, sales, transport costs, share prices, imports and exports and so on. These indexes can be useful because:

(i) They can provide background information against which businessmen can compare their performance and which provide material for government decisions.
(ii) Pay and pensions are 'indexed' by being automatically tied to the Index of Retail Prices, so that the purchasing power of money income and pension is preserved at times of rising prices.
(iii) Index numbers can be used to make projections to ascertain likely future changes in prices, output and so on. These comparisons and projections facilitate decision making at company, industry and government level.

12.2 CALCULATION OF A WEIGHTED INDEX NUMBER

In an attempt to overcome the problem of the relative difference in importance of items in an index and different units of measurement, a weighted index number makes the figures directly comparable.

The weights reflect the relative importance of an item. Therefore if food is given a weight of 300 points and housing a weight of 150 it means that a change in the price of food is twice as important as a change in the price of housing.

Given a system of weighting an index number can be calculated by the price relative method (Table 12.1):

Table 12.1 **price relative method**

Item	(1) Base price (£) (1979)	(2) Price (£) (1981)	(3) % increase	(4) Weight	(5) Product (3) × (4)
Food	25	30	20	300	6,000
Housing	20	22	10	150	1,500
Transport	5	10	100	100	10,000
Services	10	12	20	50	1,000
				600	18,500

$$\text{Index number} = \frac{18{,}500}{600} = 30.83\%$$

Therefore if the index was assumed to be 100 in 1979 (the base year) it would be 130.83 in 1981. This represents a weighted average increase in prices of 30.83%.

In the calculation (Table 12.1), the increase in price in 1981 (column 2) over 1979 (column 1) is calcualted as a percentage (column 3). This is found by taking the difference in prices over the 1979 price and multiplying by 100 to arrive at the percentage increase (or decrease).

Therefore

$$\frac{5}{25} \times 100 = 20\%$$

and

$$\frac{2}{20} \times 100 = 10\%$$

The percentage increase in price is then multiplied by the weight (column 4) to arrive at the product (column 5).

Therefore $20 \times 300 = 6000$

and $10 \times 150 = 1500$

The products are added together (18,500), and the result is divided by the sum of the weights (600):

Therefore the weighted average = $\dfrac{18{,}500}{600}$ = 30.83%

Therefore on average prices in 1981 have risen by 30.83% over prices in 1979.

If the weighting system is changed a new index (Table 12.2) can be calculated:

Table 12.2   **a new index**

Item	(1) % increase (1979-81)	(2) Weight	(3) Product (1) × (2)
Food	20	220	4,400
Housing	10	150	1,500
Transport	100	200	20,000
Services	20	30	600
		600	26,500

$$\text{Index number} = \frac{26{,}500}{600} = 44.17\% \text{ or } 144.17$$

The new weights produce a higher index number because Transport has become a more important item and has had the fastest rise in prices.

Two commonly used types of index number are:

(i) The *Laspeyre index*, which uses base-year quantities and weights. This indicates how much the cost of buying base-year quantities at current-year prices is compared with base-year costs. Different years can be directly compared with each other.

(ii) The *Paasche index*, which uses the current-year quantities and weights. This indicates how much current-year costs are related to the cost of buying current-year quantities at base-year prices. This index requires actual quantities to be ascertained for each year of the series. The different years can be compared only with the base year and not with each other.

12.3 CHAIN-BASED INDEX NUMBER

This is a system in which each period in the series uses the previous period as the base. Therefore the figures for one year (say 1980) provide the base for the next year (1981).

For example: Table 12.3.

Table 12.3 conversion from a fixed to a chain base

Year	Fixed	Chain
1978	100	100
1979	110	$\frac{110}{100} \times 100 = 110$
1980	125	$\frac{125}{110} \times 100 = 113.6$
1981	130	$\frac{130}{113.6} \times 100 = 114.4$

This is a useful system where information on the immediate past is more important than information relating to the more distant past, because it indicates the extent of change from year to year. Also this system emphasises the rate of change (in Table 12.3 the rate of change was much less from 1980 to 1981 than from 1979 to 1980).

12.4 PROBLEMS IN INDEX-NUMBER CONSTRUCTION

(a) **The choice of items** to be included in an index can present problems. If every item is included the construction of the index may become overcomplicated. Some items may have a small influence on the index. On the other hand if items are omitted then some people's interests will be ignored.

For example: the Index of Retail Prices in the UK tends to leave out very exotic and luxurious goods and people who spend their money on hanggliding in a fur coat may feel unrepresented by the index.

(b) **The choice of weights** is a considerable problem because the weights are an attempt to reflect the importance of items.

For example: the Index of Retail Prices is concerned with price changes for a 'typical household' and like an 'average man' this does not exist. Therefore the weights may not exactly reflect the importance of items for any particular household, although they may be close to a large number.

(c) **The choice of the base year** can present problems because of the 'bias' that can arise from the choice. If a 'depressed' year is chosen, the years that follow may appear particularly good; if a 'prosperous' year is chosen, the years that follow may appear poor. What is needed is a 'normal' year, which may be hard to find. At least boom and depression years can be avoided.

For example: unemployment may appear high if a year of low unemployment is chosen as the base year; and it may appear to have fallen dramatically in subsequent years if a depression year is chosen.

Also, the base year should not be too far in the past or it will be out of date and will appear unrealistic. Therefore base years are changed from time to time. These leads to a further problem.

(d) **Comparing indexes** based on different years is difficult. A change of base year brings an index up to date, but breaks the continuity of a series. This can be overcome by providing comparative figures using the old base for a time.

For example: in 1974 the Index of Retail Prices base date was changed from 16 January 1962 to 15 January 1974. The general retail price index on 19 February 1974 was 101.7 on a base of 15 January 1974 (= 100), and 195.1 on a base of 16 January 1962 (= 100). It would be possible to construct a whole new series based on the new base date:

February 1974 based on January 1974 = 101.7
February 1974 based on January 1962 = 195.1

January 1962 based on January 1974 =

$$\frac{101.7}{195.1} \times 100 = 52$$

However, comparisons of index numbers based on different years is often very difficult because of possible changes made in the construction of the numbers, changes in the weights and changes in 'expectations'. Comparisions can suffer from the problems of historical perspective.

For example: it is often stated that a particular period is more difficult for 'first-time house-buyers', because prices have risen so rapidly. Comparisons between house prices and the income of first-time house-buyers may show that the ratio has remained fairly stable. Assuming this is the case it would not be the whole story, because the number of people who want to buy a house and who expect to be able to buy a house may have increased and their expectation of the standard of house they should be able to buy may have risen.

12.5 THE INDEX OF RETAIL PRICES

This is probably the best-known index number because it is of interest to everybody as the 'cost-of-living' index. In the UK the index 'measures the change from month to month in the average level of prices of the commodities and services purchased by nearly nine-tenths of the households in the United Kingdom', (Department of Employment).

The prices of some 350 goods and services are regularly collected and approximately 150,000 separate price quotations are used each month in compiling the index. The expenditure pattern on which the index is based is revised each year using information from the Family Expenditure Survey (see Sections 2.4 and 5.15). Up to 1974 the weighting pattern was established on the base dates (June 1947, January 1952, January 1956, January 1962), but in 1974 it was decided that the weighting pattern should be revised annually in January on the basis of the information obtained from the FES.

The Retail Prices Index measures the change in the cost of a representative basket of goods and services. the composition of this basket (the relative weights attached to the various goods and services it contains) is based on the FES.

The FES is an annual survey based on a three-stage sample, with a stratified sample used at each stage (see Section 5.15). Each household has an equal chance of selection and the weights are based on the 'average' household's basket of goods. Items are given weights out of a total of 1000.

The relative importance of an item depends on how necessary it is, changes in the price of it and other commodities and changes in income. Examples of these weights are lsited in Table 12.4.

Table 12.4 **weights**

Items	1947	1952	1956	1962	1974
Food	348	399	350	314	253
Housing	88	72	87	107	124
Transport	–	–	68	100	135
Services	79	91	58	56	54

It can be seen from Table 12.4 that housing and transport have both risen in importance steadily through the years. Food has fluctuated as a result of changing prices and changes in income. As the income of households rise, the same amount is spent on food and more on other items. Services have declined in 'importance' largely because of changes in definition. New categories have been introduced into the weighting system, such as transport.

As a cost-of-living index, the Index of Retail Prices is widely quoted, although it has to be used with some caution in this context (see Sections 12.1 and 12.4). The fact that the index stood at 101.7 in February 1974 based on January 1974 indicates a rise in inflation of 1.7% in a month. In February 1980 the index stood at 248.8 on a base of January 1974 (= 100). Therefore in just over six years' inflation could be said to have risen by nearly 150%.

However, it should be clear that these figures provide no more than a general and average indication. For particular households inflation may have been greater or less than this depending on their patterns of expenditure. The index number for 'fuel and light' stood at 278.2 in March 1980 (Jan. 1974 = 100) and for 'clothing and footwear' at 199.8. Well-insulated households spending a high proportion of their income on clothes and shoes might have experienced more limited price rises than the average.

Although the Index of Retail Prices cannot provide a completely accurate and comprehensive account of inflation, it does provide the best available indicator of changes in the cost of living and the level of inflation in the UK.

ASSIGNMENTS

1 Compare the latest available figure for the Index of Retail Prices and the index five years ago. How good is the index as an indication of changes in the cost of living?

Table 12.5 index of industrial production

	Seasonally adjusted 1975 = 100						By market sector		
	All industries covered §	By industry					Consumer goods industries	Investment goods industries	Intermediate goods industries
		Mining and quarrying	Manufacturing	Construction[2]	Gas, electricity and water				
1975 weights	1000	41	697	182	80		243	218	349
1969	99.6	123.9	97.6	113.5	80.9		89.4	95.7	102.9
1970	99.7	119.1	98.0	111.4	84.1		91.0	95.6	102.9
1971	99.8	119.1	97.4	113.3	87.3		93.3	93.7	102.1
1972	102.0	100.2	100.0	115.4	93.6		99.9	91.2	104.1
1973	109.5	110.1	108.4	118.2	99.3		108.0	99.0	112.6
1974	105.1	89.9	106.5	105.8	99.2		106.1	102.1	106.1
1975	100.0	100.0	100.0	100.0	100.0		100.0	100.0	100.0
1976	102.0	125.8	101.4	98.6	102.9		101.9	96.8	107.1
1977	106.0	187.7	103.1	98.3	107.1		104.7	98.6	115.5
1978	109.9	232.5	103.8	105.6	110.2		107.0	98.5	121.4
1979	112.7	293.5	104.2	102.8	117.1		106.3	99.9	130.4

Source: *Economic Trends* (April 1980) table 26.

2 Write a report on the UK Indexes of Distribution.

3 Table 12.5 shows the UK Index of Industrial Production. Write a commentary in about 1000 words on the information contained in this table.

4 A firm uses four raw materials in its production. Calculate a weighted index number from the information in the table below:

Raw materials	Base price 1979 (£)	New price 1981 (£)	Weight
A	35	40	253
B	28	35	124
C	8	20	135
D	12	16	54

Write a report on the results of the calculations.

5 Discuss the problems involved in constructing an index number.

6 Wages, pensions, prices and savings can be linked to an index. Consider the advantages and problems involved in index-linking.

CHAPTER 13
COMPARING STATISTICS: CORRELATION

13.1 WHAT IS CORRELATION?

Correlation is concerned with whether or not there is any association between two variables. If two variables are related to any extent, then changes in the value of one are associated with changes in the value of the other.

For example: an increase in the sales of a company may show a strong association with increases in the money spent on the advertising of its products.

Connections between two variables are an everyday occurrence (for instance: the income and expenditure of a household, the costs and sales of a company, the miles driven in a car and the petrol bought for it).

It is useful to know:
(i) Whether any association exists.
(ii) The strength of the association.
(iii) The direction of the relationship.
(iv) The proportion of the variability in one variable that can be accounted for by its relationship with the other variable.

It is useful to try to measure these factors because:

(i) Knowledge of the relationships enables plans and predictions to be made.
(ii) Past evidence can be used to make decisions.
(iii) Plans and predictions based on evidence rather than guesswork provide greater control over events.

For example: if it is known how many kilometres or miles a commercial vehicle travels using a litre or a gallon of petrol, it is possible to predict how much petrol will be required for a particular journey.

Also, if it is assumed that the kilometres travelled by a vehicle and the petrol consumed are strongly related, then a sudden change in petrol consumption may provide an early indicator of a problem in the engine.

If it is assumed that a particular vehicle always consumes 5 litres (just over a gallon) of petrol to cover 50 kilometres (just over 30 miles), then the association would appear on a graph as in Figure 13.1.

Fig 13.1 *perfect correlation*

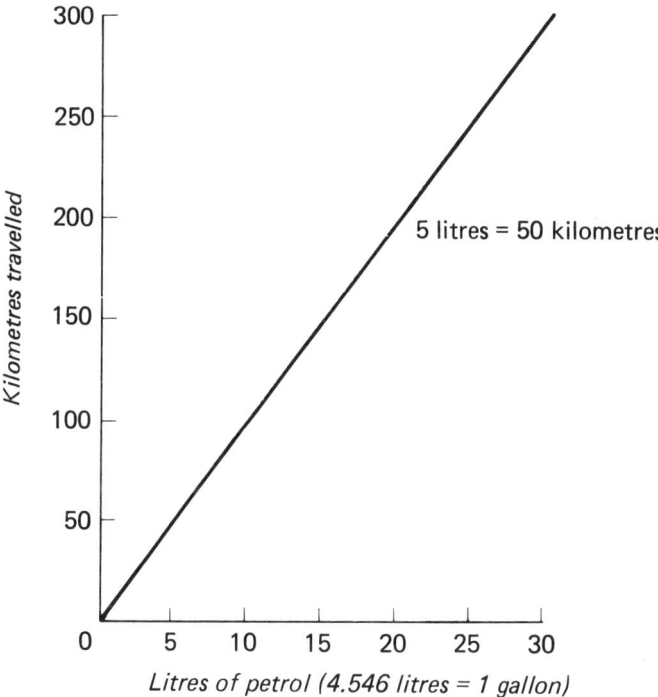

Figure 13.1 is an example of perfect correlation or association. The fact that the 'curve' is a straight line makes it possible to read off one variable against the other (see Section 10.8).

For example: if a journey was to be undertaken of 220 kilometres, it would be clear that the vehicle would consume 22 litres of petrol. If there were 12 litres of petrol in the vehicle it would be known that it could travel exactly 120 kilometres on this amount of petrol.

In fact many relationships between two variables are linear (that is, based on a straight line), but very few are as perfect as this example. There are many non-linear relationships as well and the calculation of the correlation coefficient in these cases is complicated. In considering correlation, trends and time series, the main concern here is with linear relationships.

Few linear relationships are perfect because other factors are involved.

For example: if a company spends more on advertising, this does not guarantee a rise in sales and certainly any rise in sales that does take place will not be by a predetermined amount. Competition from other companies, the economic situation, the success with which the advertising is carried out, may all influence the connection between these two variables.

There are a number of methods of indicating correlation between two variables. These include:

(i) Scatter diagrams.
(ii) Correlation tables.
(iii) The product moment coefficient of correlation.
(iv) The coefficient of rank correlation.
(v) Regression.

13.2 SCATTER DIAGRAMS

Scatter diagrams (or scatter graphs) provide a useful means of deciding whether or not there is association between variables.

The construction of a scatter diagram is by drawing a graph so that the scale for one variable (the independent variable, if this can be determined) lies along the horizontal axis, and the other variable (the dependent variable) on the vertical axis. Each pair of figures is then plotted as a single point on the graph.

The basic purpose is to see whether there is any pattern among the points. If there is, a 'line of best fit' can be drawn with the same number of points or co-ordinates on each side of the line. This line is the one judged to be the best line to fit the pattern of points.

For example: Figure 13.2.

In Figure 13.2 the correlation indicated by the line of best fit is approximately linear. Also it indicates positive correlation, that is as advertising expenditure is increased, so sales revenue tends to rise.

This is not 'proved' by this graph, nor is any causal relationship clear from it. All that can be said with certainty is that there appears to be some association between the two variables. The scatter diagram provides an indication which may be the first piece of evidence in an investigation. A scatter diagram can indicate the type of correlation that exists (see Figures 13.3, 13.4, 13.5, 13.6 and 13.7):

Fig 13.2 *scatter diagrams: advertising expenditure and sales revenue of a company*

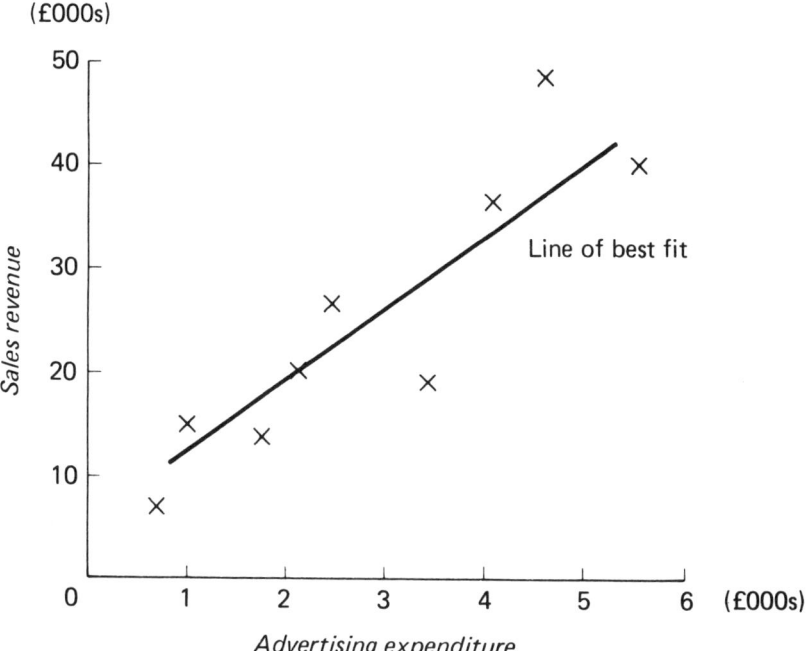

Fig 13.3 *positive correlation*

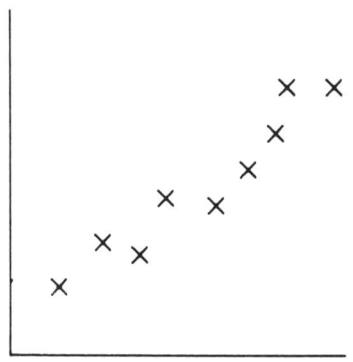

Fig 13.4 *negative correlation*

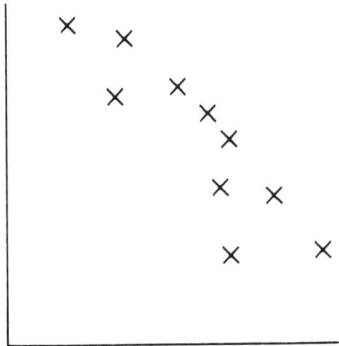

Fig 13.5 *perfect positive correlation*

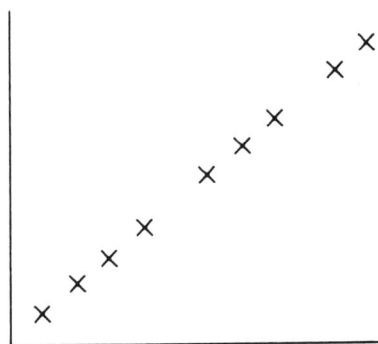

Fig 13.6 *perfect negative correlation*

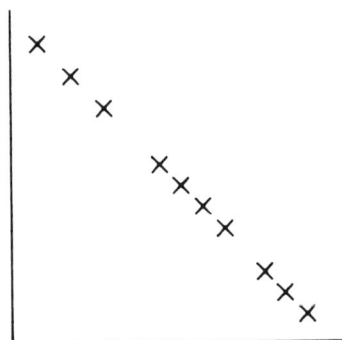

Fig 13.7 *no correlation*

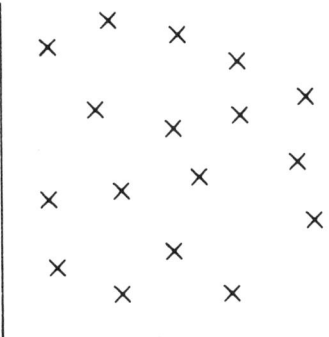

(i) Figure 13.3 indicates positive correlation so that as the variable x increases so y will increase.
(ii) Figure 13.4 indicates negative correlation so that as x increases y will decrease.
(iii) Figure 13.5 indicates perfect positive correlation between the two variables so that they both increase in the same proportions.
(iv) Figure 13.6 indicates perfect negative correlation between the two variables so that as one rises the other falls in exact proportion.
(v) Figure 13.7 indicates that there is no correlation between the two variables. There are a number of lines of best fit which could be drawn with equal validity.

Once the line of best fit has been drawn a scatter diagram can be used for estimating by reading off the line the value of one variable against the other. However, this is not likely to be very reliable unless the two variables are very closely related.

Scatter diagrams are limited because:

(i) There is uncertainty as to the correct position of the line of best fit.
(ii) Although there is an indication of the type of correlation, there is not any indication of the extent of the correlation.

13.3 CORRELATION TABLES

These perform a similar function to scatter diagrams. They provide an approximate indication of correlation.

For example: Table 13.1

Table 13.1 **a correlation table**

Age of employees (to the nearest year)	Length of service (to the nearest year)						
	Under 15	5-9	10-14	15-19	20-24	25-29	30 and above
16-25	14	7					
26-35	1	18					
36-45		2	28	18	1		
46-55			10	22	36		
56-65		1			24	12	6

Table 13.1 shows a comparison between the age of a company's employees and their length of service. As would be expected, the older the employees the longer their service has been with the company. In other words there is an indication of positive correlation. The table shows, for example:

(i) That 110 out of 200 employees are over 46 years old (55%) and therefore that the company has a fairly old labour force.
(ii) Only 20% of the labour force are below the age of 36.
(iii) Nobody over the age of 36 has been newly employed by the company for at least five years.

Although this type of information can be seen from the table, the information it provides on correlation is limited.

13.4 THE PRODUCT MOMENT COEFFICIENT OF CORRELATION

The 'product moment' coefficient is a measure of linear correlation (devised by Karl Pearson and sometimes referred to as Pearson's coefficient), which attempts to show the closeness of the relationship between two variables. Therefore this is not only an attempt to show whether correlation exists (which is also the aim of scatter diagrams and correlation tables), but also an attempt to show how closely it exists.

The coefficient is based on a formula which divides the mean product of the deviations from the mean (shown by $\Sigma(xy)$), by the product of the standard deviations (shown by $\sigma_x \; \sigma_y$):

$$r = \frac{\Sigma(xy)}{n\sigma_x\sigma_y}$$

where r is the product moment coefficient
x is one variable (which is sometimes referred to as the subject)
y is the other variable (which can be referred to as the relative)
n is the number of items (subject or relative, not both)
σ_x is the standard deviation of variable x
σ_y is the standard deviation of variable y

This formula can be rewritten on the basis of the basic formula for the standard deviation:

$$r = \frac{\Sigma(xy)}{\sqrt{\Sigma x^2 \Sigma y^2}}$$

where $\Sigma(xy)$ represents the mean product of the deviations from the mean
x is the deviation from the mean of one variable
y is the deviation from the mean of the other variable

For example: Table 13.2.

Table 13.2 **product moment coefficient**

Year	Investment (£000s)	Profit (£000s)	x (deviation from $\bar{x}$ of investment)	y (deviation from $\bar{x}$ of profit)	x^2	y^2	xy
1978	3	8	−2	−2	4	4	4
1979	4	5	−1	−5	1	25	5
1980	6	7	+1	−3	1	9	3
1981	7	20	+2	+10	4	100	20
		20	20	40		10	138	32

$\bar{x}$ (Investment) $= \dfrac{20}{4} = 5$ $\Sigma(xy) = 32$, $\Sigma x^2 = 10$, $\Sigma y^2 = 138$

$\bar{x}$ (Profit) $= \dfrac{40}{4} = 10$ $r = \dfrac{32}{\sqrt{10 \times 138}} = \dfrac{32}{37.15} = +0.86.$

In Table 13.2 the coefficient (+ 0.86) shows a strong positive correlation between the two variables.

13.5 THE MEANING OF r

(a) **Direction**: the sign + or − indicates the direction of association (positive or negative) between the two variables.

(b) **Strength**: the numerical value of r indicates the strength of the linear relationship. r can only lie between −1 and +1: +0.9 indicates strong positive correlation, −0.9 indicates strong negative correlation, and + or − 0.1 indicates weak correlation. Below about 0.7 correlation is not very strong (r^2 would be 0.49, see (c) below). The importance of differences between say 0.9 and 0.8 should not be over-emphasised. All that can be said is that they both indicate strong correlation.

(c) **Variability**: r is the proportion of the variability in one variable that can be accounted for by its linear relationship with the other variable. Therefore, if $r = +0.8$ then $r^2 = +0.64$. It can be said that 64% of the variability in one variable (say x) can be accounted for by its linear relationship with the other variable (y).

(d) **Reliability**: when the number of observations is very small it is not possible to place any reliance on the value of r, because any apparent relationship between the two variables may be by chance or coincidence. With a large number of observations this will be less likely.

(e) **Lag**: although strong correlation may be shown between two variables, there may be a lapse of time (a time lag) before one item influences the other. For instance advertising may not have an immediate effect on sales. This time lag may give an impression of negative correlation, when in fact correlation would be positive if due allowance was made for lag.

13.6 COEFFICIENT OF RANK CORRELATION

Some information is provided only on an ordinal scale; other information may be on an interval scale but is ordered or ranked.

For example: the capacity of containers such as bottles and boxes may be known, but they may be listed in size order from the smallest to the largest without the detail of their capacity included.

In these cases it is possible to calculate a coefficient of rank correlation (r'). The interpretation of r' is the same as for r.

To calculate the coefficient of rank correlation:

(i) Rank each respective variable in order.
(ii) Put the corresponding rank of one variable against the rank of the other variable.
(iii) Show the difference between the rankings in each case.
(iv) Square the differences and find the sum of the squares.
(v) Apply the formula:

$$r' = 1 - \frac{6\Sigma d^2}{n(n^2 - 1)}$$

For example: test results on a commodity range for the effects of weathering are produced in ranked order. These results are compared with the predicted rankings based on the judgement of a group of experts. Both rankings are shown in Table 13.3.

Table 13.3 **rank correlation coefficient**

Commodity	(1) Predicted rankings	(2) Test rankings	(3) d (2) − (1)	(4) d^2 $(3)^2$
A	1	5	4	16
B	2	3	1	1
C	3	4	1	1
D	4	2	2	4
E	5	1	4	16
				38

$\Sigma d^2 = 38$

In the formula

$$r' = 1 - \frac{6\Sigma d^2}{n(n^2 - 1)}$$

d^2 = the square of the differences or deviations in the rankings
n = the number of units

Therefore applied to Table 13.3:

$$r' = 1 - \frac{6 \times 38}{5(25 - 1)}$$

$$= 1 - \frac{228}{5 \times 24}$$

$$= 1 - \frac{228}{120}$$

$$= 1 - 1.9$$

$$= \underline{-0.9}$$

This indicates a high level of negative correlation between the tests and the predicted results.

13.7 SPURIOUS CORRELATION

(a) **Causal relationships**: it cannot be over-emphasised that because two variables are associated and may produce a high correlation coefficient, this does not prove that there is a causal relationship between them. Correlation does not indicate cause and effect.

For example: there may be a strong positive correlation between sales revenue and advertising expenditure over a given period of time. It may be assumed that high sales have been caused by greater advertising expenditure. However, this is not the only possible interpretation. It could be that in years when sales were high and the company was prosperous that a higher advertising budget was allocated. Or possibly, both higher sales and greater advertising expenditure arose from a general expansion in the economy.

(b) **Indirect connections**: there may be no direct connection between two variables that produce a high correlation coefficient.

For example: sales of computers may show a high positive correlation with the level of television viewing. There may be some connection between them because both computers and television can be described as being part of the electronics industry, but the question in correlation is whether one directly influences the other. It seems unlikely that an increasing sale of computers will increase television viewing directly, or that if television viewing was reduced that this would necessarily reduce the sale of computers. Therefore it may be that although the correlation coefficient between the two variables is high, there is in fact very limited association between them.

(c) **Coincidence**: it is possible to want to be able to predict the occurrence of a variable and therefore to search amongst a mass of data to find anything which correlates highly with this variable.

For example: at various times in history economic recession has been blamed on 'sunspots', because it has been noticed that these have appeared during recessions. Therefore it has been argued that there is a strong positive correlation between the appearance of sunspots and the probability of recession. It can be argued that sunspots may be connected in some way with adverse weather conditions causing poor harvests and therefore encouraging recession. However, it is generally felt (by economists and agricultural experts) that this has not been well established and is an example of spurious correlation.

(d) *In general*: correlation is no more than an indicator of possible association between two variables. It is one piece of evidence which can lead to further investigation. Therefore it can be seen as a first stage. If little association is indicated by a correlation coefficient or a scatter diagram, then there is little point in applying the next technique, which is regression.

13.8 REGRESSION

Regression attempts to show the relationship between two variables by providing a mean line which best indicates the trend of the points or co-ordinates on a graph.

Correlation coefficients do not provide any information on the slope of the line between two variables. The slope indicates the rate of change in one variable against the other. Regression lines do show this slope.

The line of best fit on a scatter diagram is dependent on the subjective judgement of the person who draws it. A regression line is drawn mathematically and therefore it is independent of individual judgement. The aim is to minimise the total divergence of the points or co-ordinates from the line. Mathematically it has been found that the best line is one that minimises the total of the squared deviations. This is known as the method of least squares.

The method of calculating a regression line by least squares is described in Section 14.5. These aspects of correlation and regression have been included in Chapter 14 because of the close links with charting time series and trends.

ASSIGNMENTS

1

Year	Sales (£000)	Advertising cost £00
1976	10	2
1977	15	3
1978	12	5
1979	18	10
1981	20	12

Calculate the product moment coefficient and a rank correlation coefficient for the above data.

Comment on the meaning of the results.

What are the disadvantages of using a rank correlation coefficient as compared with the product moment coefficient?

2 What are the problems of interpreting the results of the calculation of a correlation coefficient?

3 Draw scatter diagrams for the information in Table 13.2 and Table 13.3. Comment on the results.

4 Discuss the meaning of the term correlation. Comment on the various methods of indicating correlation.

5 What is meant by 'spurious correlation'? Discuss the extent to which there is a connection between spurious correlation and spurious accuracy.

CHAPTER 14

TRENDS AND FORECASTING

14.1 TRENDS AND FORECASTING

Forecasts are based on information about the way in which variables have been behaving in the past. In forecasting it has to be assumed that the behavioural patterns that have been traced in the past will continue in the future for a reasonable time.

What is reasonable depends on the variable. National economic prosperity or depression is likely to change only over a period of years; while the sale of ice-creasem may change weekly or daily, depending on the weather. Yet even with ice-cream there are trends in sales which can be seen over much longer periods such as seasons, or over a period of years.

Statistical projections do not necessarily produce accurate forecasts, because any analysis of trends depends on the assumption of stable political, economic and social conditions. However, trends can provide the possibility of:

(a) **Control**: for planning it is useful to monitor what has happened and to know directly circumstances change. In business and for management purposes the analysis of trends are used for such areas as budgetary control, stock-holding, investment planning and market research.

National governments have similar interests in planning and control to monitor changes in productivity, the balance of payments, levels of unemployment and levels of inflation. Economic indicators, such as the Index of Retail Prices, are carefully watched for short- and long-term developments.

(b) **Interpolation**: this involves finding a value within the past trend which may be a useful guide to future action.

(c) **Extrapolation**: extending a trend into the future may provide an indication of what is likely to happen. However, this process of extrapolation always requires great caution because the variables themselves may change, circumstances may alter and the estimated trend may not be very accurate.

Statistical techniques are useful in forecasting. Averages can provide a starting-point, dispersion can show how representative an average is, correlation can indicate the level of association between sets of data and index numbers can be used to compare present and past performance.

What is required also is an analysis of data over time in a more general application than that supplied by index numbers. This analysis is provided by a time series.

14.2 TIME SERIES

A time series consists of numerical data recorded at intervals of time.

For example: weekly output, monthly sales, annual profit.

A time series is a special case of the two-variable situation found in correlation, in which one variable is always time. Time is the independent variable (graphed on the horizontal or x axis) because it changes at regular intervals (weeks, months, quarters, years). A graph is the most popular form of presentation because this most clearly illustrates the relationship between the dependent variable and time, as well as clearly showing the trend.

There are a variety of statistical techniques aimed at separating the various elements of the trend. Some of the basic techniques are covered in this chapter. The objective of analysing the trend is to learn about the behaviour of the series and to use this knowledge as a basis for future action.

Most time series can be separated into fairly clear types of trend:

(i) the long-term trend (also called the basic or secular trend);
(ii) cyclical fluctuations, which are superimposed on the long-term trend;
(iii) seasonal variations (or short-period movements);
(iv) irregular fluctuations (or non-recurring, random, spasmodic or residual fluctuations).

The analysis of time series is aimed at disentangling these fluctuations.

For example: the long-term trend is for prices to rise, but there are cyclical fluctuations during which prices may fall, or at least rise more slowly than at other times. On a seasonal basis, prices tend to rise in the period leading up to Christmas. Irregular fluctuations in prices may occur because of particular shortages caused by strikes, or poor harvests, and because of sudden changes in such areas as taxation.

The techniques used to analyse these trends include:

(i) Moving averages, to show the long-term trend.
(ii) Seasonal adjustments, to provide control for seasonal variations.
(iii) Straight-line graphs to emphasise the general direction of the trend and to remove short-term and cyclical fluctuations.

14.3 MOVING AVERAGES

Moving averages are a method of repeatedly calculating a series of different average values along a time series to produce a trend line. The line produced by charting a moving average on a graph is not a straight line, but it does even out short-term and cyclical fluctuations to some extent.

Some figures will be above the average, others below it; by using an average, fluctuations are offset one against another to produce the trend. Normally the length of the moving average (two-year, five-year, etc.) will be based on the period of time (weeks, months, years) between successive peaks or successive troughs.

For example: Table 14.1

Table 14.1 moving average: annual sales of Company *A* over a fifteen-year period

Year	Sales (£m.)		5-year moving average
1968	5		
1969	2		
1970	4	$32 \div 5 = 6.4$	6.4
1971	9		8.8
1972	12		10.4
1973	17		10.8
1974	10		11.6
1975	6		13.2
1976	13		13.4
1977	20		13.2
1978	18		15.2
1979	9		17.0
1980	16		17.0
1981	22		
1982	20		

Table 14.1 shows the annual sales of company with a five-year moving average. This moving average is calculate by:

(i) Looking at the sales and noticing that the intervals between high points and the intervals between low points are about five years. Therefore a 5-year moving average is calculated.
(ii) Adding up the sales figures for the first five years and dividing by five to produce an average ($32 \div 5 = 6.4$).
(iii) Subtracting the first year and adding the sixth year to obtain the next moving average figure ($-5 + 17 = 32 + 12 = 44 \div 5 = 8.8$), and then continuing this through the series.
(iv) Plotting the moving average trend line against the sales figures, as shown in Figure 14.1.

Fig 14.1 *moving average: annual sales of Company* A

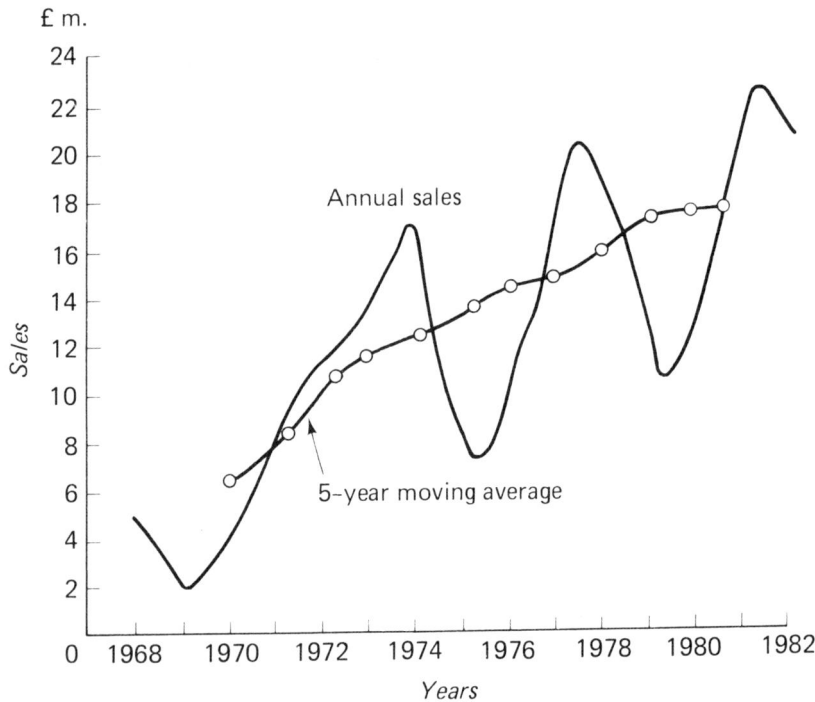

From Figure 14.1 it is clear that:

(i) The long-term trend of sales is upwards.
(ii) There are cyclical fluctuations at fairly regular intervals.

The moving average evens out the cyclical fluctuations and provides a clear indication of the long-term trend. More data would be required to extend the moving average across the whole series.

14.4 SEASONAL VARIATIONS

The effects of seasonal variations are often eliminated from data in order to clarify the underlying trend.

For example: in many government publications time series are produced with a qualification added to the table or graph that the figures are 'seasonally adjusted'. For instance, unemployment figures are 'seasonally adjusted', which is defined as 'adjusted for normal seasonal variations'. Seasonal variations in unemployment include rising unemployment in the winter due to the weather, and rapid changes in the summer months due to school and student leavers joining the unemployment register at the end of the academic year.

There are a number of methods of adjusting for seasonal variations. Table 14.2 shows a commonly used method which illustrates the general principles involved:

Table 14.2 **seasonal adjustments**

Year and quarters		(1) Sales (£m.)	(2) 4-quarterly totals	(3) Centred totals	(4) Trend	(5) Variations from the trend
1978	1	11				
	2	8	50			
	3	13	54	104	13	0
	4	18	54	108	13.5	+4.5
1979	1	15	60	114	14.3	+0.7
	2	9	66	126	15.8	−6.8
	3	18	67	133	16.6	+1.4
	4	24	69	136	17	+7

Year	Q	(1)	(2)	(3)	(4)	(5)
1980	1	16	76	145	18.1	−2.1
	2	11	80	156	19.5	−8.5
	3	25	82	162	20.3	−4.7
	4	28	83	165	20.6	−7.4
1981	1	18	82	165	20.6	−2.6
	2	12	84	166	20.8	−8.8
	3	24				
	4	30				

In Table 14.2:

(i) Column (1) shows the sales in £m.
(ii) Column (2) shows the 4-quarterly totals (11 +8 + 13 + 18 = 50, 8 + 13 + 18 + 15 = 54, etc.).
(iii) Column (3) shows the total of each pair of 4-quarterly totals (50 + 54 = 104, 54 + 54 = 108, etc.).
(iv) Column (4) shows the trend, arrived at by dividing the centred total in column (3) by 8 (104 ÷ 8 = 13, 108 ÷ 8 = 13.5, etc.).
(v) Column (5) shows the variations between sales in column (1) and the trend in column (4) (13 − 13 = 0, 18 − 13.5 = + 4.5, etc.).

It is from column (5) of Table 14.2 that the seasonal variations are calculated (Table 14.3):

Table 14.3 **seasonal variations**

	Year	\multicolumn{4}{c}{Quarters}

	Year	1	2	3	4
	1978	−	−	0	+4.5
	1979	+0.7	−6.8	+1.4	+7
	1980	−2.1	−8.5	+4.7	+7.4
	1981	−2.6	−8.8		
	Totals	−4	−24.1	+6.1	+18.9
(1)	Average	−1.3	−8.03	+2.03	+6.3
(2)	Adjustment	+0.25	+0.25	+0.25	+0.25
(3)	Seasonal variation	−1.1	−7.8	+2.3	+6.6

In Table 14.3:

(i) The average (1) is the total for each of the quarters divided by the appropriate number of quarters: $+0.7 - 2.1 - 2.6 = -4 \div 3 = -1.3$.
(ii) The adjustment (2) is based on the 'unexplained' deviations that have not been entirely eliminated. If they had, the sum of the averages would have been zero: $-1.3 - 8.03 + 2.03 + 6.3 = -9.33 + 8.33 = -1$.
(iii) This excess negative figure is taken from each quarter equally by dividing by 4 ($1 \div 4 = 0.25$) and adding this figure (0.25) to each quarterly average. It is added to each quarterly figure because it is not clear where the difference arises. In this case the difference is very small, but it can be large in some cases.
(iv) The final row (3) is the estimate of seasonal variations for each quarter. The figures are rounded to one decimal place, because to extend the figures further would indicate a higher degree of accuracy than is likely.

These figures can be used to produce a seasonally adjusted series and they can be applied to future years.

For example: Table 14.4 and Figure 14.2.

Table 14.4 seasonally adjusted sales figures: 1982 sales figures for Company B

Quarters	Sales (£m.)	Seasonal adjustment	Seasonally adjusted sales (£m.)
1	20	+1.1	21.1
2	15	+7.8	22.8
3	25	−2.3	22.7
4	35	−6.6	28.4

Notice that in Table 14.4 the minus figures for seasonal variation have been added and the plus figures subtracted. What is being done is to eliminate seasonal variation and therefore additions are made to the generally low seasons, and subtractions made to the seasons in which the variable has a high value.

It is clear from Figure 14.2 that when seasonal variations are eliminated the general trend of sales is upwards although not quite as sharply as indicated by the fourth quarter's unadjusted results would indicate. The fall in sales in the second quarter appears to be because of seasonal factors.

Fig 14.2 *seasonally adjusted sales figure: 1982 sales figures for Company B*

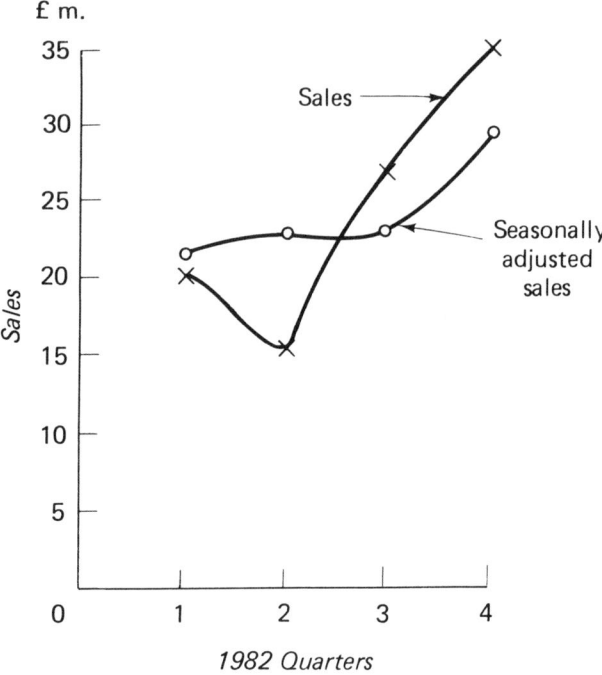

These points are borne out if the whole series is drawn on a time series graph (Table 14.5 and Figure 14.3):

Table 14.5 **time series with seasonal adjustments**

Year and quarter		Sales (£m.)	Seasonally adjusted sales (£m.)
1978	1	11	12.1
	2	8	15.8
	3	13	10.7
	4	18	11.4
1979	1	15	16.1
	2	9	16.8
	3	18	15.7
	4	24	17.4

1980	1	16	17.1
	2	11	18.8
	3	25	22.7
	4	28	21.4
1981	1	18	19.1
	2	12	19.8
	3	24	21.7
	4	30	23.4

Fig 14.3 *time series with seasonal adjustments*

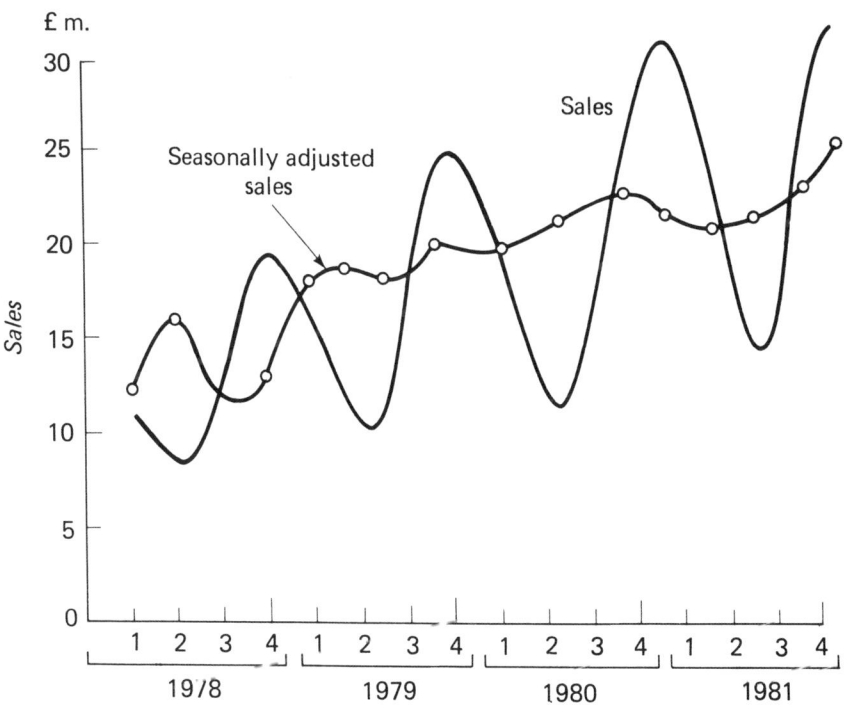

14.5 IRREGULAR AND RESIDUAL FLUCTUATIONS

In a time series the curve of the graph reflects the trend, seasonal variations and irregular factors. The value of the residuals can be calculated once the trend and the seasonal variations have been calculated (Table 14.6):

Table 14.6 **residual factors**

Years and quarters		Original series	=	Trend	+	Seasonal variation	+	Residual
1978	1	11						
	2	8						
	3	13		13		+2.3		−2.3
	4	18		13.5		+6.6		−2.1
1979	1	15		14.3		−1.1		+1.8
	2	9		15.8		−7.8		+1
	3	18		16.6		+2.3		−0.9
	4	24		17		+6.6		+0.4
1980	1	16		18.1		−1.1		−1
	2	11		19.5		−7.8		−0.7
	3	25		20.3		+2.3		+2.4
	4	28		20.6		+6.6		+0.8
1981	1	18		20.6		−1.1		−1.5
	2	12		20.8		−7.8		−1
	3	24						
	4	30						

In Table 14.6 a residual figure has been calculated in the last column. The value of calculating this figure is to give an indication of the extent to which the series has been affected by irregular external factors. The residual values calculated in Table 14.6 are small and therefore any forecast made on the basis of the time series would be unlikely to be upset by external factors. If the residual factors had been large then any forecasts made would be less reliable.

14.6 LINEAR TRENDS

There are many methods of arriving at a linear or straight line trend. Perhaps the two most straightforward methods are:

(a) three-point linear regression,
(b) the least-squares method.

(a) Three-point linear regression.
This is also known as the method of partial averages, or semi-averages, or three-point estimation.

The straight line is found by:

(i) Calculating the arithmetic mean of the series.
(ii) Calculating the arithmetic mean of the figures which are positioned above point (i).
(iii) Calculating the arithmetic mean of the figures which are positioned below point (i).
(iv) Plotting the three points on a graph to form a straight line.

The sales figures in Table 14.7 can be graphed along with the five-year moving average and the three-point linear regression line (Figure 14.4).

For example: Table 14.7.

Table 14.7 three-point linear regression

Year	Sales (£m.)	
1968	5	
1969	2	
1970	4	
1971	9	$\frac{59}{7} = 8.4$
1972	12	
1973	17	
1974	10	
1975	6	Arithmetic mean = $\frac{183}{15}$ = 12.2
1976	13	
1977	20	
1978	18	
1979	9	$\frac{118}{7} = 16.9$
1980	16	
1981	22	
1982	20	

183

Notice that in Figure 14.4 the first point (i) is placed at the centre of the time series (1975). If there were two numerical values (such as sales against advertising expenditure) the arithmetic mean of both series would be calculated and plotted against each other. This divides both series into two parts and the points (ii) and (iii) can then be calculated.

Fig 14.4 *three-point linear regression*

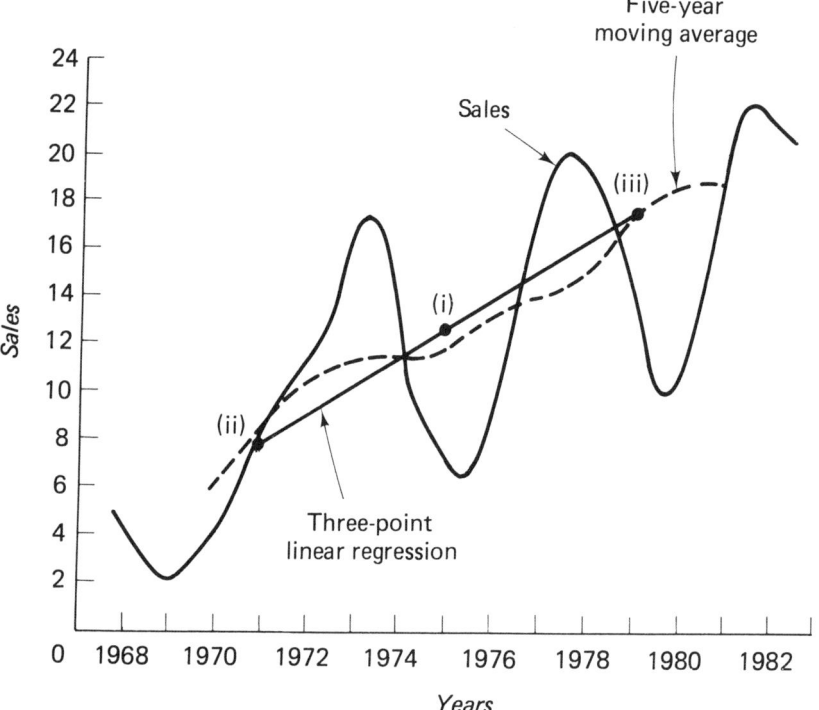

The straight line approximates closely to the five-year moving average. Clearly, it smooths out cyclical fluctuations and provides a clear picture of the basic trend. It would be possible to extrapolate this line into the future; however, given the cyclical fluctuations in the basic series and other possible factors, any forecasts would have to be very cautious. Also in general this method is applicable only when the trend is linear or approximately linear. Therefore, while this method can provide a clear overall impression, it should not be used indiscriminately.

(b) The least-squares method

This method also produces a linear trend line. The aim is to produce a line which minimises all positive and negative deviations of the data from a straight line (the 'best' line of best fit) drawn through the data. This is carried out by squaring the deviations (to remove minus items, in the same way as they were removed for the standard deviation), and therefore the least squares is the 'best' line the line which minimises the error in the direction of the variable being predicted.

The equation for finding a straight line is:

$$y = a + bx$$

where y and x are the two variables, a represents the intercept and b represents the slope of the line.

The 'intercept' is the distance between the origin and the point where the straight line meets the vertical axis.

The 'slope' of the line (as indicated in Section 13.8) shows changes in one variable against the other. This is illustrated in Figure 14.5.

Fig 14.5 *the slope and the intercept*

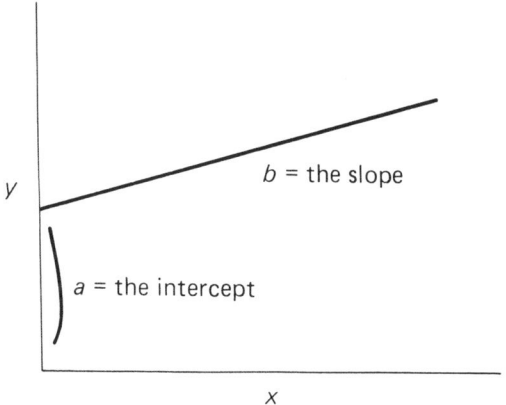

The straight-line equation $y = a + bx$ can be rewritten as:

$$y = \bar{y} - b\bar{x} + bx$$

where $\bar{y}$ and $\bar{x}$ are the arithmetic mean of the two variables. This equation, in the same way as the previous one, minimises the positive and negative deviations of the data from a straight line trend.

In a time series one of the two sets of data (x) is time and this increases by equal amounts (months or years). This time series does not have a 'value', but it is a series so that it is possible to number the years (months, week or days) in order as an increase of one on the last (0, 1, 2, 3, etc.). When the variable x is not time, then this line is a regression line.

An example of the calculation of the least-squares method is shown in Table 14.8.

Table 14.8 the least-squares method

Year x	Sales (£m.) y	x^2	y^2	xy
1968 (0)	5	0	25	0
1969 (1)	2	1	4	2
1970 (2)	4	4	16	8
1971 (3)	9	9	81	27
1972 (4)	12	16	144	48
1973 (5)	17	25	289	85
1974 (6)	10	36	100	60
1975 (7)	6	49	36	42
1976 (8)	13	64	169	114
1977 (9)	20	81	400	180
1978 (10)	18	100	324	180
1979 (11)	9	121	81	99
1980 (12)	16	144	256	192
1981 (13)	22	169	484	286
1982 (14)	20	196	400	280
105	183	1015	2809	1603

In Table 14.8:

$$\bar{x} = \frac{105}{15} = 7 \qquad \bar{y} = \frac{183}{15} = 12.2$$

$$b = \frac{\Sigma xy - \dfrac{\Sigma x \times \Sigma y}{n}}{\Sigma x^2 - \dfrac{(\Sigma x)^2}{n}}$$

$$= \frac{1603 - \frac{105 \times 183}{15}}{1015 - \frac{105^2}{15}}$$

$$= \frac{1603 - 1281}{1015 - 735}$$

$$= \frac{322}{280}$$

$$= 1.15$$

$$y = a + bx \quad \text{or} \quad y = \bar{y} - b\bar{x} + bx$$
$$y = 12.2 - (1.15 \times 7) + (1.15 \times x)$$
$$= 12.2 - 8.05 + 1.15x$$
$$= 4.15 + 1.15x$$

Therefore $y = 4.15 + 1.15x$.

This means that each year (x) there will be an average increase of £1.15 (million) in sales. When $x = 5$, y will equal $4.15 + 1.15 \times 5 = 9.90$.

If this is applied to the annual sales of Company A, the results in Table 14.9 are produced:

Table 14.9 **linear trend**

Year		Sales (£m.)	Linear trend
1968	(0)	5	$y = 4.15 + 1.15 \times 0 = 4.15$
1969	(1)	2	$y = 4.15 + 1.15 \times 1 = 5.3$
1970	(2)	4	$y = 4.15 + 1.15 \times 2 = 6.45$
1971	(3)	9	$y = 4.15 + 1.15 \times 3 = 7.60$
1972	(4)	12	$y = 4.15 + 1.15 \times 4 = 8.75$
1973	(5)	17	$y = 4.15 + 1.15 \times 5 = 9.90$
1974	(6)	10	$y = 4.15 + 1.15 \times 6 = 11.05$
1975	(7)	6	$y = 4.15 + 1.15 \times 7 = 12.20$
1976	(8)	13	$y = 4.15 + 1.15 \times 8 = 13.35$
1977	(9)	20	$y = 4.15 + 1.15 \times 9 = 14.50$
1978	(10)	18	$y = 4.15 + 1.15 \times 10 = 15.65$
1979	(11)	9	$y = 4.15 + 1.15 \times 11 = 16.80$
1980	(12)	16	$y = 4.15 + 1.15 \times 12 = 17.95$
1981	(13)	22	$y = 4.15 + 1.15 \times 13 = 19.10$
1982	(14)	20	$y = 4.15 + 1.15 \times 14 = 19.25$

The linear (least-squares) trend in Table 14.9 can be graphed against sales (Figure 14.6):

Fig 14.6 *least-squares trend: annual sales of Company A*

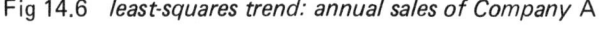

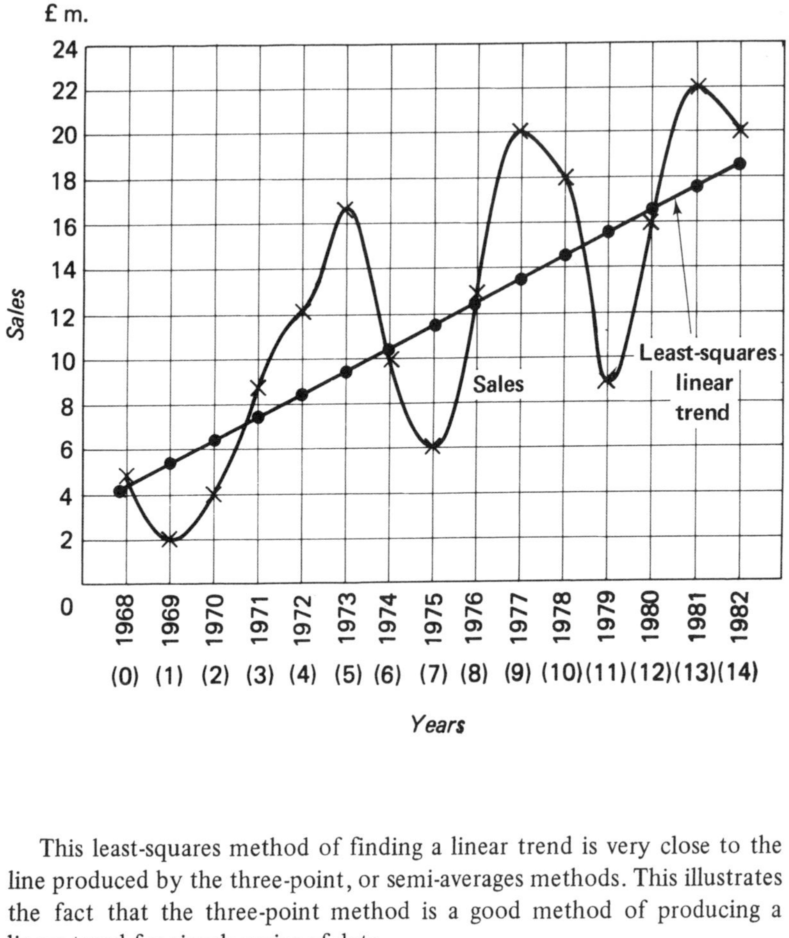

This least-squares method of finding a linear trend is very close to the line produced by the three-point, or semi-averages methods. This illustrates the fact that the three-point method is a good method of producing a linear trend for simple series of data.

Both methods are based on averaging out deviations. In both cases the straight line suggests a limited set of influences acting together in a single direction. The straight line smooths out short-term and cyclical fluctuations to emphasise the basic trend.

14.7 CONCLUSIONS

The methods for finding trends described in this chapter are only some of the many available. This is an area in which there has been a good deal of research and development because of the desire by business and government to be able to forecast. Analysing the various types of trend can indicate a pattern in the data which may be repeated in the future. However, this analysis is not a substitute for looking carefully at the original figures and attempting to analyse why poor results occurred in some years and good results in others. It may be possible to avoid repeating the poor results and to try to ensure the repetition of the good results.

Any forecasts that are made need to be surrounded with qualifications because of all the unexpected and unforeseen variables that can influence future performance. It is still well worth forecasting because planning can provide control over events and avoid constant last-minute problem solving.

Statistics can provide the basis for planning, decision making and forecasting. The methods and concepts outlined in this book are aimed at providing a framework which can be used for understanding, collecting, presenting, summarising, comparing and interpreting statistics. It is only by 'handling' statistics, that is with practice, that confidence in the ability to use these techniques can be created.

14.8 THE NEXT STEP

The aim of this book is to provide a comprehensive introduction to statistics, and appreciation of the subject and a foundation for further study. It is not a mathematics book, it is intended to show the practical applications of statistical techniques. One area of further study is a 'step back' to look at the proof of the formulas and the mathematical basis for the techniques used here. This would require a greater mathematical and algebraic knowledge than is assumed in this book.

For the lay reader, the social scientist, the businessman and the business studies student the next step is to go further along the statistical road in one or all of four main areas:

(a) **Sources of data**: there is a wide area of further reading and study in the sources mentioned here. In all regions and countries and in all types of business, economic and social activity there are a range of statistical sources worth further investigation.

(b) **The collection of data**: there are a number of areas of survey methods and sampling in which a very detailed study is possible, linked to practical experience in collecting data.

(c) **Statistical methods**: it has been suggested in various chapters that there are many more complex statistical techniques than those discussed in this book, although most of them are based on the techniques outlined here.

(d) **Applications**: statistics can be applied to a variety of subjects and problems, many of which have been suggested in this book. These applications include using statistics in sales, market research, forecasting, personnel work, aspects of production, computing, business decision-making, solving business problems as well as in the development of economics, sociology, psychology and many other subject areas.

The reading list (Appendix A.1) is intended to provide suggestions for complimentary and further reading.

ASSIGNMENT

1 Company sales (£000s)

	1	2	3	4
1978			120	100
1979	85	140	130	90
1980	90	120	125	85
1981	110	125	140	95
1982	115	130		

Show the trend for the above time series by calculating a moving average.

Plot the sales and the moving average on a graph.

Calculate the seasonal variations for these figures and show the seasonally adjusted sales curve on the graph.

Comment on your results and on the 3 curves on the graph.

2 Calculate a three-point linear trend line from the weekly wage figures in Table 10.1. Calculate a trend line by the least-squares method.

Draw a graph showing the two lines.

Comment on the validity of the methods of producing a linear trend and on the information brought out by the lines.

3 Discuss the reasons for attempting to forecast trends and the problems of statistical forecasting.

4 Find details of forecasting made by the government, companies and other organisations. Consider the basis on which they have been made and, if possible, how accurate they have proved to be.

5 Consider in detail the problems of weather forecasting. How applicable are these problems to the problems of forecasting faced by business?

APPENDIXES

A.1 READING LIST

(a) **Statistical sources**
Central Statistical Office, *Guide to Official Statistics* (HMSO, 1978).
Central Statistical Office, *Social Trends* (HMSO, annually).
Central Statistical Office, *Annual Abstract of Statistics* (HMSO, annually).
Central Statistical Office, *Economic Trends* (HMSO, monthly).
Central Statistical Office, *Statistical News* (HMSO, quarterly).

(b) **Collection of data**
C. A. Moser and G. Kalton, *Survey Methods in Social Investigation*, 2nd rev. ed. (London: Heinemann, 1971).
A. N. Oppenheim, *Questionnaire Design and Attitude Measurement*, new ed. (London: Heinemann, 1968).

(c) **Statistical methods**
B. Edwards, *The Readable Maths and Statistics Book* (London: Allen & Unwin, 1980).
M. R. Spiegel, *Theory and Problems of Statistics*, Schaum's Outline Series (New York: McGraw-Hill, 1961).
H. M. Blalock, *Social Statistics* (New York: McGraw-Hill, 1960).
P. Montagon, *Foundations of Statistics: A Survey for Managers* (Cheltenham: Thornes, 1980).

(d) **Application of statistics**
C. Robson, *Experiment, Design and Statistics in Psychology* (Harmondsworth: Penguin, 1973).
A. E. Maxwell, *Basic Statistics in Behavioural Research* (Harmondsworth: Penguin, 1970).
W. J. Reichmann, *Use and Abuse of Statistics* (Harmondsworth: Penguin, 1964).
D. Huff, *How to Lie with Statistics* (Harmondsworth: Penguin, 1973).

A.2 AREA TABLE

Areas under the normal curve

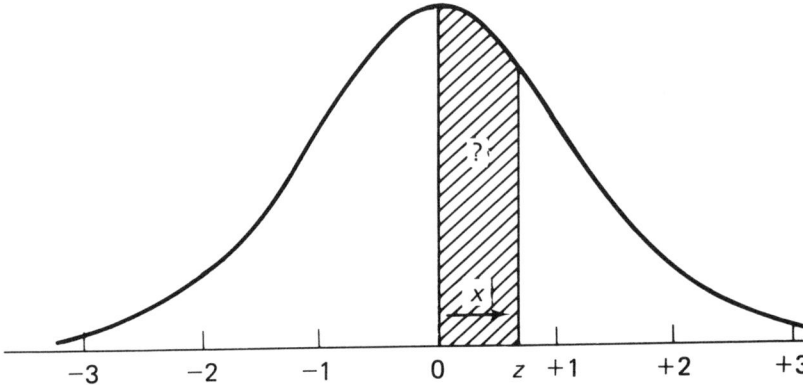

x is the distance from the mean measured in standard deviations to arrive at z values:

$$z = \frac{x - \bar{x}}{\sigma}$$

z	Area	z	Area
0.0	0.0000	1.5	0.4332
0.1	0.0398	1.6	0.4452
0.2	0.0793	1.7	0.4554
0.3	0.1179	1.8	0.4641
0.4	0.1554	1.9	0.4713
0.5	0.1915	2.0	0.4772
0.6	0.2257	2.1	0.4821
0.7	0.2580	2.2	0.4861
0.8	0.2881	2.3	0.4893
0.9	0.3159	2.4	0.4918
1.0	0.3413	2.5	0.4938
1.1	0.3643	2.6	0.4953
1.2	0.3849	2.7	0.4965
1.3	0.4032	2.8	0.4974
1.4	0.4192	2.9	0.4981
		3.0	0.4987

A.3 SQUARE ROOT TABLES

Table A

SQUARE ROOTS. From 1 to 10

	0	1	2	3	4	5	6	7	8	9	Mean Differences								
											1	2	3	4	5	6	7	8	9
1·0	1·000	1·005	1·010	1·015	1·020	1·025	1·030	1·034	1·039	1·044	0	1	1	2	2	3	3	4	4
1·1	1·049	1·054	1·058	1·063	1·068	1·072	1·077	1·082	1·086	1·091	0	1	1	2	2	3	3	4	4
1·2	1·095	1·100	1·105	1·109	1·114	1·118	1·122	1·127	1·131	1·136	0	1	1	2	2	3	3	4	4
1·3	1·140	1·145	1·149	1·153	1·158	1·162	1·166	1·170	1·175	1·179	0	1	1	2	2	3	3	3	4
1·4	1·183	1·187	1·192	1·196	1·200	1·204	1·208	1·212	1·217	1·221	0	1	1	2	2	2	3	3	4
1·5	1·225	1·229	1·233	1·237	1·241	1·245	1·249	1·253	1·257	1·261	0	1	1	2	2	2	3	3	4
1·6	1·265	1·269	1·273	1·277	1·281	1·285	1·288	1·292	1·296	1·300	0	1	1	2	2	2	3	3	3
1·7	1·304	1·308	1·311	1·315	1·319	1·323	1·327	1·330	1·334	1·338	0	1	1	2	2	2	3	3	3
1·8	1·342	1·345	1·349	1·353	1·356	1·360	1·364	1·367	1·371	1·375	0	1	1	1	2	2	3	3	3
1·9	1·378	1·382	1·386	1·389	1·393	1·396	1·400	1·404	1·407	1·411	0	1	1	1	2	2	3	3	3
2·0	1·414	1·418	1·421	1·425	1·428	1·432	1·435	1·439	1·442	1·446	0	1	1	1	2	2	2	3	3
2·1	1·449	1·453	1·456	1·459	1·463	1·466	1·470	1·473	1·476	1·480	0	1	1	1	2	2	2	3	3
2·2	1·483	1·487	1·490	1·493	1·497	1·500	1·503	1·507	1·510	1·513	0	1	1	1	2	2	2	3	3
2·3	1·517	1·520	1·523	1·526	1·530	1·533	1·536	1·539	1·543	1·546	0	1	1	1	2	2	2	3	3
2·4	1·549	1·552	1·556	1·559	1·562	1·565	1·568	1·572	1·575	1·578	0	1	1	1	2	2	2	3	3
2·5	1·581	1·584	1·587	1·591	1·594	1·597	1·600	1·603	1·606	1·609	0	1	1	1	2	2	2	3	3
2·6	1·612	1·616	1·619	1·622	1·625	1·628	1·631	1·634	1·637	1·640	0	1	1	1	2	2	2	2	3
2·7	1·643	1·646	1·649	1·652	1·655	1·658	1·661	1·664	1·667	1·670	0	1	1	1	2	2	2	2	3
2·8	1·673	1·676	1·679	1·682	1·685	1·688	1·691	1·694	1·697	1·700	0	1	1	1	1	2	2	2	3
2·9	1·703	1·706	1·709	1·712	1·715	1·718	1·720	1·723	1·726	1·729	0	1	1	1	1	2	2	2	3
3·0	1·732	1·735	1·738	1·741	1·744	1·746	1·749	1·752	1·755	1·758	0	1	1	1	1	2	2	2	3
3·1	1·761	1·764	1·766	1·769	1·772	1·775	1·778	1·780	1·783	1·786	0	1	1	1	1	2	2	2	3
3·2	1·789	1·792	1·794	1·797	1·800	1·803	1·806	1·808	1·811	1·814	0	1	1	1	1	2	2	2	2
3·3	1·817	1·819	1·822	1·825	1·828	1·830	1·833	1·836	1·838	1·841	0	1	1	1	1	2	2	2	2
3·4	1·844	1·847	1·849	1·852	1·855	1·857	1·860	1·863	1·865	1·868	0	1	1	1	1	2	2	2	2
3·5	1·871	1·873	1·876	1·879	1·881	1·884	1·887	1·889	1·892	1·895	0	1	1	1	1	2	2	2	2
3·6	1·897	1·900	1·903	1·905	1·908	1·910	1·913	1·916	1·918	1·921	0	1	1	1	1	2	2	2	2
3·7	1·924	1·926	1·929	1·931	1·934	1·936	1·939	1·942	1·944	1·947	0	1	1	1	1	2	2	2	2
3·8	1·949	1·952	1·954	1·957	1·960	1·962	1·965	1·967	1·970	1·972	0	1	1	1	1	2	2	2	2
3·9	1·975	1·977	1·980	1·982	1·985	1·987	1·990	1·992	1·995	1·997	0	1	1	1	1	2	2	2	2
4·0	2·000	2·002	2·005	2·007	2·010	2·012	2·015	2·017	2·020	2·022	0	0	1	1	1	1	2	2	2
4·1	2·025	2·027	2·030	2·032	2·035	2·037	2·040	2·042	2·045	2·047	0	0	1	1	1	1	2	2	2
4·2	2·049	2·052	2·054	2·057	2·059	2·062	2·064	2·066	2·069	2·071	0	0	1	1	1	1	2	2	2
4·3	2·074	2·076	2·078	2·081	2·083	2·086	2·088	2·090	2·093	2·095	0	0	1	1	1	1	2	2	2
4·4	2·098	2·100	2·102	2·105	2·107	2·110	2·112	2·114	2·117	2·119	0	0	1	1	1	1	2	2	2
4·5	2·121	2·124	2·126	2·128	2·131	2·133	2·135	2·138	2·140	2·142	0	0	1	1	1	1	2	2	2
4·6	2·145	2·147	2·149	2·152	2·154	2·156	2·159	2·161	2·163	2·166	0	0	1	1	1	1	2	2	2
4·7	2·168	2·170	2·173	2·175	2·177	2·179	2·182	2·184	2·186	2·189	0	0	1	1	1	1	2	2	2
4·8	2·191	2·193	2·195	2·198	2·200	2·202	2·205	2·207	2·209	2·211	0	0	1	1	1	1	2	2	2
4·9	2·214	2·216	2·218	2·220	2·223	2·225	2·227	2·229	2·232	2·234	0	0	1	1	1	1	2	2	2
5·0	2·236	2·238	2·241	2·243	2·245	2·247	2·249	2·252	2·254	2·256	0	0	1	1	1	1	2	2	2
5·1	2·258	2·261	2·263	2·265	2·267	2·269	2·272	2·274	2·276	2·278	0	0	1	1	1	1	2	2	2
5·2	2·280	2·283	2·285	2·287	2·289	2·291	2·293	2·296	2·298	2·300	0	0	1	1	1	1	2	2	2
5·3	2·302	2·304	2·307	2·309	2·311	2·313	2·315	2·317	2·319	2·322	0	0	1	1	1	1	2	2	2
5·4	2·324	2·326	2·328	2·330	2·332	2·335	2·337	2·339	2·341	2·343	0	0	1	1	1	1	1	2	2

Table B

SQUARE ROOTS. FROM 1 TO 10.

	0	1	2	3	4	5	6	7	8	9	Mean Differences.								
											1	2	3	4	5	6	7	8	9
5·5	2·345	2·347	2·349	2·352	2·354	2·356	2·358	2·360	2·362	2·364	0	0	1	1	1	1	1	2	2
5·6	2·366	2·369	2·371	2·373	2·375	2·377	2·379	2·381	2·383	2·385	0	0	1	1	1	1	1	2	2
5·7	2·387	2·390	2·392	2·394	2·396	2·398	2·400	2·402	2·404	2·406	0	0	1	1	1	1	1	2	2
5·8	2·408	2·410	2·412	2·415	2·417	2·419	2·421	2·423	2·425	2·427	0	0	1	1	1	1	1	2	2
5·9	2·429	2·431	2·433	2·435	2·437	2·439	2·441	2·443	2·445	2·447	0	0	1	1	1	1	1	2	2
6·0	2·449	2·452	2·454	2·456	2·458	2·460	2·462	2·464	2·466	2·468	0	0	1	1	1	1	1	2	2
6·1	2·470	2·472	2·474	2·476	2·478	2·480	2·482	2·484	2·486	2·488	0	0	1	1	1	1	1	2	2
6·2	2·4900	2·492	2·494	2·496	2·498	2·500	2·502	2·504	2·506	2·508	0	0	1	1	1	1	1	2	2
6·3	2·510	2·512	2·514	2·516	2·518	2·520	2·522	2·524	2·526	2·528	0	0	1	1	1	1	1	2	2
6·4	2·530	2·532	2·534	2·536	2·538	2·540	2·542	2·544	2·546	2·548	0	0	1	1	1	1	1	2	2
6·5	2·550	2·551	2·553	2·555	2·557	2·559	2·561	2·563	2·565	2·567	0	0	1	1	1	1	1	2	2
6·6	2·569	2·571	2·573	2·575	2·577	2·579	2·581	2·583	2·585	2·587	0	0	1	1	1	1	1	2	2
6·7	2·588	2·590	2·592	2·594	2·596	2·598	2·600	2·602	2·604	2·606	0	0	1	1	1	1	1	2	2
6·8	2·608	2·610	2·612	2·613	2·615	2·617	2·619	2·621	2·623	2·625	0	0	1	1	1	1	1	2	2
6·9	2·627	2·629	2·631	2·632	2·634	2·636	2·638	2·640	2·642	2·644	0	0	1	1	1	1	1	2	2
7·0	2·646	2·648	2·650	2·651	2·653	2·655	2·657	2·659	2·661	2·663	0	0	1	1	1	1	1	2	2
7·1	2·665	2·666	2·668	2·670	2·672	2·674	2·676	2·678	2·680	2·681	0	0	1	1	1	1	1	1	2
7·2	2·683	2·685	2·687	2·689	2·691	2·693	2·694	2·696	2·698	2·700	0	0	1	1	1	1	1	1	2
7·3	2·702	2·704	2·706	2·707	2·709	2·711	2·713	2·715	2·717	2·718	0	0	1	1	1	1	1	1	2
7·4	2·720	2·722	2·724	2·726	2·728	2·729	2·731	2·733	2·735	2·737	0	0	1	1	1	1	1	1	2
7·5	2·739	2·740	2·742	2·744	2·746	2·748	2·750	2·751	2·753	2·755	0	0	1	1	1	1	1	1	2
7·6	2·757	2·759	2·760	2·762	2·764	2·766	2·768	2·769	2·771	2·773	0	0	1	1	1	1	1	1	2
7·7	2·775	2·777	2·778	2·780	2·782	2·784	2·786	2·787	2·789	2·791	0	0	1	1	1	1	1	1	2
7·8	2·793	2·795	2·796	2·798	2·800	2·802	2·804	2·805	2·807	2·809	0	0	1	1	1	1	1	1	2
7·9	2·811	2·812	2·814	2·816	2·818	2·820	2·821	2·823	2·825	2·827	0	0	1	1	1	1	1	1	2
8·0	2·828	2·830	2·832	2·834	2·835	2·837	2·839	2·841	2·843	2·844	0	0	1	1	1	1	1	1	2
8·1	2·846	2·848	2·850	2·851	2·853	2·855	2·857	2·858	2·860	2·862	0	0	1	1	1	1	1	1	2
8·2	2·864	2·865	2·867	2·869	2·871	2·872	2·874	2·876	2·877	2·879	0	0	1	1	1	1	1	1	2
8·3	2·881	2·883	2·884	2·886	2·888	2·890	2·891	2·893	2·895	2·897	0	0	1	1	1	1	1	1	2
8·4	2·898	2·900	2·902	2·903	2·905	2·907	2·909	2·910	2·912	2·914	0	0	1	1	1	1	1	1	2
8·5	2·915	2·917	2·919	2·921	2·922	2·924	2·926	2·927	2·929	2·931	0	0	1	1	1	1	1	1	2
8·6	2·933	2·934	2·936	2·938	2·939	2·941	2·943	2·944	2·946	2·948	0	0	1	1	1	1	1	1	2
8·7	2·950	2·951	2·953	2·955	2·956	2·958	2·960	2·961	2·963	2·965	0	0	1	1	1	1	1	1	2
8·8	2·966	2·968	2·970	2·972	2·973	2·975	2·977	2·978	2·980	2·982	0	0	1	1	1	1	1	1	2
8·9	2·983	2·985	2·987	2·988	2·990	2·992	2·993	2·995	2·997	2·998	0	0	1	1	1	1	1	1	2
9·0	3·000	3·002	3·003	3·005	3·007	3·008	3·010	3·012	3·013	3·015	0	0	0	1	1	1	1	1	1
9·1	3·017	3·018	3·020	3·022	3·023	3·025	3·027	3·028	3·030	3·032	0	0	0	1	1	1	1	1	1
9·2	3·033	3·035	3·036	3·038	3·040	3·041	3·043	3·045	3·046	3·048	0	0	0	1	1	1	1	1	1
9·3	3·050	3·051	3·053	3·055	3·056	3·058	3·059	3·061	3·063	3·064	0	0	0	1	1	1	1	1	1
9·4	3·066	3·068	3·069	3·071	3·072	3·074	3·076	3·077	3·079	3·081	0	0	0	1	1	1	1	1	1
9·5	3·082	3·084	3·085	3·087	3·089	3·090	3·092	3·094	3·095	3·097	0	0	0	1	1	1	1	1	1
9·6	3·098	3·100	3·102	3·103	3·105	3·106	3·108	3·110	3·111	3·113	0	0	0	1	1	1	1	1	1
9·7	3·114	3·116	3·118	3·119	3·121	3·122	3·124	3·126	3·127	3·129	0	0	0	1	1	1	1	1	1
9·8	3·130	3·132	3·134	3·135	3·137	3·138	3·140	3·142	3·143	3·145	0	0	0	1	1	1	1	1	1
9·9	3·146	3·148	3·150	3·151	3·153	3·154	3·156	3·158	3·159	3·161	0	0	0	1	1	1	1	1	1

Table C

SQUARE ROOTS. FROM 10 TO 100

	0	1	2	3	4	5	6	7	8	9	Mean Differences 1 2 3	4 5 6	7 8 9
10	3·162	3·178	3·194	3·209	3·225	3·240	3·256	3·271	3·286	3·302	2 3 5	6 8 9	11 12 14
11	3·317	3·332	3·347	3·362	3·376	3·391	3·406	3·421	3·435	3·450	1 3 4	6 7 9	10 12 13
12	3·464	3·479	3·493	3·507	3·521	3·536	3·550	3·564	3·578	3·592	1 3 4	6 7 8	10 11 13
13	3·606	3·619	3·633	3·647	3·661	3·674	3·688	3·701	3·715	3·728	1 3 4	5 7 8	10 11 12
14	3·742	3·755	3·768	3·782	3·795	3·808	3·821	3·834	3·847	3·860	1 3 4	5 7 8	9 11 12
15	3·873	3·886	3·899	3·912	3·924	3·937	3·950	3·962	3·975	3·987	1 3 4	5 6 8	9 10 11
16	4·000	4·012	4·025	4·037	4·050	4·062	4·074	4·087	4·099	4·111	1 2 4	5 6 7	9 10 11
17	4·123	4·135	4·147	4·159	4·171	4·183	4·195	4·207	4·219	4·231	1 2 4	5 6 7	8 10 11
18	4·243	4·254	4·266	4·278	4·290	4·301	4·313	4·324	4·336	4·347	1 2 3	5 6 7	8 9 10
19	4·359	4·370	4·382	4·393	4·405	4·416	4·427	4·438	4·450	4·461	1 2 3	5 6 7	8 9 10
20	4·472	4·483	4·494	4·506	4·517	4·528	4·539	4·550	4·561	4·572	1 2 3	4 6 7	8 9 10
21	4·583	4·593	4·604	4·615	4·626	4·637	4·648	4·658	4·669	4·680	1 2 3	4 5 6	8 9 10
22	4·690	4·701	4·712	4·722	4·733	4·743	4·754	4·764	4·775	4·785	1 2 3	4 5 6	7 8 9
23	4·796	4·806	4·817	4·827	4·837	4·848	4·858	4·868	4·879	4·889	1 2 3	4 5 6	7 8 9
24	4·899	4·909	4·919	4·930	4·940	4·950	4·960	4·970	4·980	4·990	1 2 3	4 5 6	7 8 9
25	5·000	5·010	5·020	5·030	5·040	5·050	5·060	5·070	5·079	5·089	1 2 3	4 5 6	7 8 9
26	5·099	5·109	5·119	5·128	5·138	5·148	5·158	5·167	5·177	5·187	1 2 3	4 5 6	7 8 9
27	5·196	5·206	5·215	5·225	5·235	5·244	5·254	5·263	5·273	5·282	1 2 3	4 5 6	7 8 9
28	5·292	5·301	5·310	5·320	5·329	5·339	5·348	5·357	5·367	5·376	1 2 3	4 5 6	7 7 8
29	5·385	5·394	5·404	5·413	5·422	5·431	5·441	5·450	5·459	5·468	1 2 3	4 5 5	6 7 8
30	5·477	5·486	5·495	5·505	5·514	5·523	5·532	5·541	5·550	5·559	1 2 3	4 4 5	6 7 8
31	5·568	5·577	5·586	5·595	5·604	5·612	5·621	5·630	5·639	5·648	1 2 3	3 4 5	6 7 8
32	5·657	5·666	5·675	5·683	5·692	5·701	5·710	5·718	5·727	5·736	1 2 3	3 4 5	6 7 8
33	5·745	5·753	5·762	5·771	5·779	5·788	5·797	5·805	5·814	5·822	1 2 3	3 4 5	6 7 8
34	5·831	5·840	5·848	5·857	5·865	5·874	5·882	5·891	5·899	5·908	1 2 3	3 4 5	6 7 8
35	5·916	5·925	5·933	5·941	5·950	5·958	5·967	5·975	5·983	5·992	1 2 2	3 4 5	6 7 8
36	6·000	6·008	6·017	6·025	6·033	6·042	6·050	6·058	6·066	6·075	1 2 2	3 4 5	6 7 7
37	6·083	6·091	6·099	6·107	6·116	6·124	6·132	6·140	6·148	6·156	1 2 2	3 4 5	6 7 7
38	6·164	6·173	6·181	6·189	6·197	6·205	6·213	6·221	6·229	6·237	1 2 2	3 4 5	6 6 7
39	6·245	6·253	6·261	6·269	6·277	6·285	6·293	6·301	6·309	6·317	1 2 2	3 4 5	6 6 7
40	6·325	6·332	6·340	6·348	6·356	6·364	6·372	6·380	6·387	6·395	1 2 2	3 4 5	6 6 7
41	6·403	6·411	6·419	6·427	6·434	6·442	6·450	6·458	6·465	6·473	1 2 2	3 4 5	5 6 7
42	6·481	6·488	6·496	6·504	6·512	6·519	6·527	6·535	6·542	6·550	1 2 2	3 4 5	5 6 7
43	6·557	6·565	6·573	6·580	6·588	6·595	6·603	6·611	6·618	6·626	1 2 2	3 4 5	5 6 7
44	6·633	6·641	6·648	6·656	6·663	6·671	6·678	6·686	6·693	6·701	1 2 2	3 4 5	5 6 7
45	6·708	6·716	6·723	6·731	6·738	6·745	6·753	6·760	6·768	6·775	1 1 2	3 4 4	5 6 7
46	6·782	6·790	6·797	6·804	6·812	6·819	6·826	6·834	6·841	6·848	1 1 2	3 4 4	5 6 7
47	6·856	6·863	6·870	6·877	6·885	6·892	6·899	6·907	6·914	6·921	1 1 2	3 4 4	5 6 7
48	6·928	6·935	6·943	6·950	6·957	6·964	6·971	6·979	6·986	6·993	1 1 2	3 4 4	5 6 6
49	7·000	7·007	7·014	7·021	7·029	7·036	7·043	7·050	7·057	7·064	1 1 2	3 4 4	5 6 6
50	7·071	7·078	7·085	7·092	7·099	7·106	7·113	7·120	7·127	7·134	1 1 2	3 4 4	5 6 6
51	7·141	7·148	7·155	7·162	7·169	7·176	7·183	7·190	7·197	7·204	1 1 2	3 4 4	5 6 6
52	7·211	7·218	7·225	7·232	7·239	7·246	7·253	7·259	7·266	7·273	1 1 2	3 3 4	5 6 6
53	7·280	7·287	7·294	7·301	7·308	7·314	7·321	7·328	7·335	7·342	1 1 2	3 3 4	5 5 6
54	7·348	7·355	7·362	7·369	7·376	7·382	7·389	7·396	7·403	7·409	1 1 2	3 3 4	5 5 6

Table D

SQUARE ROOTS. FROM 10 TO 100

	0	1	2	3	4	5	6	7	8	9	Mean Differences		
											1 2 3	4 5 6	7 8 9
55	7·416	7·423	7·430	7·436	7·443	7·450	7·457	7·463	7·470	7·477	1 1 2	3 3 4	5 5 6
56	7·483	7·490	7·497	7·503	7·510	7·517	7·523	7·530	7·537	7·543	1 1 2	3 3 4	5 5 6
57	7·550	7·556	7·563	7·570	7·576	7·583	7·589	7·596	7·603	7·609	1 1 2	3 3 4	5 5 6
58	7·616	7·622	7·629	7·635	7·642	7·649	7·655	7·662	7·668	7·675	1 1 2	3 3 4	5 5 6
59	7·681	7·688	7·694	7·701	7·707	7·714	7·720	7·727	7·733	7·740	1 1 2	3 3 4	4 5 6
60	7·746	7·752	7·759	7·765	7·772	7·778	7·785	7·791	7·797	7·804	1 1 2	3 3 4	4 5 6
61	7·810	7·817	7·823	7·829	7·836	7·842	7·849	7·855	7·861	7·868	1 1 2	3 3 4	4 5 6
62	7·874	7·880	7·887	7·893	7·899	7·906	7·912	7·918	7·925	7·931	1 1 2	3 3 4	4 5 6
63	7·937	7·944	7·950	7·956	7·962	7·969	7·975	7·981	7·987	7·994	1 1 2	3 3 4	4 5 6
64	8·000	8·006	8·012	8·019	8·025	8·031	8·037	8·044	8·050	8·056	0 1 2	2 3 4	4 5 6
65	8·062	8·068	8·075	8·081	8·087	8·093	8·099	8·106	8·112	8·118	1 1 2	2 3 4	4 5 6
66	8·124	8·130	8·136	8·142	8·149	8·155	8·161	8·167	8·173	8·179	1 1 2	2 3 4	4 5 5
67	8·185	8·191	8·198	8·204	8·210	8·216	8·222	8·228	8·234	8·240	1 1 2	2 3 4	4 5 5
68	8·246	8·252	8·258	8·264	8·270	8·276	8·283	8·289	8·295	8·301	1 1 2	2 3 4	4 5 5
69	8·307	8·313	8·319	8·325	8·331	8·337	8·343	8·349	8·355	8·361	1 1 2	2 3 4	4 5 5
70	8·367	8·373	8·379	8·385	8·390	8·396	8·402	8·408	8·414	8·420	1 1 2	2 3 4	4 5 5
71	8·426	8·432	8·438	8·444	8·450	8·456	8·462	8·468	8·473	8·479	1 1 2	2 3 4	4 5 5
72	8·485	8·491	8·497	8·503	8·509	8·515	8·521	8·526	8·532	8·538	1 1 2	2 3 3	4 5 5
73	8·544	8·550	8·556	8·562	8·567	8·573	8·579	8·585	8·591	8·597	1 1 2	2 3 3	4 5 5
74	8·602	8·608	8·614	8·620	8·626	8·631	8·637	8·643	8·649	8·654	1 1 2	2 3 3	4 5 5
75	8·660	8·666	8·672	8·678	8·683	8·689	8·695	8·701	8·706	8·712	1 1 2	2 3 3	4 5 5
76	8·718	8·724	8·729	8·735	8·741	8·746	8·752	8·758	8·764	8·769	1 1 2	2 3 3	4 5 5
77	8·775	8·781	8·786	8·792	8·798	8·803	8·809	8·815	8·820	8·826	1 1 2	2 3 3	4 4 5
78	8·832	8·837	8·843	8·849	8·854	8·860	8·866	8·871	8·877	8·883	1 1 2	2 3 3	4 4 5
79	8·888	8·894	8·899	8·905	8·911	8·916	8·922	8·927	8·933	8·939	1 1 2	2 3 3	4 4 5
80	8·944	8·950	8·955	8·961	8·967	8·972	8·978	8·983	8·989	8·994	1 1 2	2 3 3	4 4 5
81	9·000	9·006	9·011	9·017	9·022	9·028	9·033	9·039	9·044	9·050	1 1 2	2 3 3	4 4 5
82	9·055	9·061	9·066	9·072	9·077	9·083	9·088	9·094	9·099	9·105	1 1 2	2 3 3	4 4 5
83	9·110	9·116	9·121	9·127	9·132	9·138	9·143	9·149	9·154	9·160	1 1 2	2 3 3	4 4 5
84	9·165	9·171	9·176	9·182	9·187	9·192	9·198	9·203	9·209	9·214	1 1 2	2 3 3	4 4 5
85	9·220	9·225	9·230	9·236	9·241	9·247	9·252	9·257	9·263	9·268	1 1 2	2 3 3	4 4 5
86	9·274	9·279	9·284	9·290	9·295	9·301	9·306	9·311	9·317	9·322	1 1 2	2 3 3	4 4 5
87	9·327	9·333	9·338	9·343	9·349	9·354	9·359	9·365	9·370	9·375	1 1 2	2 3 3	4 4 5
88	9·381	9·386	9·391	9·397	9·402	9·407	9·413	9·418	9·423	9·429	1 1 2	2 3 3	4 4 5
89	9·434	9·439	9·445	9·450	9·455	9·460	9·466	9·471	9·476	9·482	1 1 2	2 3 3	4 4 5
90	9·487	9·492	9·497	9·503	9·508	9·513	9·518	9·524	9·529	9·534	1 1 2	2 3 3	4 4 5
91	9·539	9·545	9·550	9·555	9·560	9·566	9·571	9·576	9·581	9·586	1 1 2	2 3 3	4 4 5
92	9·592	9·597	9·602	9·607	9·612	9·618	9·623	9·628	9·633	9·638	1 1 2	2 3 3	4 4 5
93	9·644	9·649	9·654	9·659	9·664	9·670	9·675	9·680	9·685	9·690	1 1 2	2 3 3	4 4 5
94	9·695	9·701	9·706	9·711	9·716	9·721	9·726	9·731	9·737	9·742	1 1 2	2 3 3	4 4 5
95	9·747	9·752	9·757	9·762	9·767	9·772	9·778	9·783	9·788	9·793	1 1 2	2 3 3	4 4 5
96	9·798	9·803	9·808	9·813	9·818	9·823	9·829	9·834	9·839	9·844	1 1 2	2 3 3	4 4 5
97	9·849	9·854	9·859	9·864	9·869	9·874	9·879	9·884	9·889	9·894	1 1 1	2 3 3	4 4 5
98	9·899	9·905	9·910	9·915	9·920	9·925	9·930	9·935	9·940	9·945	0 1 1	2 2 3	3 4 4
99	9·950	9·955	9·960	9·965	9·970	9·975	9·980	9·985	9·990	9·995	0 1 1	2 2 3	3 4 4

A.4 LOGARITHM AND ANTILOGARITHM TABLES

Table A

LOGARITHMS

	0	1	2	3	4	5	6	7	8	9	1 2 3	4 5 6	7 8 9
10	0000	0043	0086	0128	0170	0212	0253	0294	0334	0374	4 9 13 4 8 12	17 21 26 16 20 24	30 34 38 28 32 36
11	0414	0453	0492	0531	0569	0607	0645	0682	0719	0755	4 8 12 4 7 11	15 19 23 15 19 22	27 31 35 26 30 33
12	0792	0828	0864	0899	0934	0969	1004	1038	1072	1106	3 7 11 3 7 10	14 18 21 14 17 20	25 28 32 24 27 31
13	1139	1173	1206	1239	1271	1303	1335	1367	1399	1430	3 7 10 3 7 10	13 16 20 13 16 19	23 26 30 22 25 29
14	1461	1492	1523	1553	1584	1614	1644	1673	1703	1732	3 6 9 3 6 9	12 15 19 12 15 17	22 25 28 20 23 26
15	1761	1790	1818	1847	1875	1903	1931	1959	1987	2014	3 6 9 3 6 8	11 14 17 11 14 17	20 23 26 19 22 24
16	2041	2068	2095	2122	2148	2175	2201	2227	2253	2279	3 5 8 3 5 8	11 14 16 10 13 16	19 22 24 18 21 23
17	2304	2330	2355	2380	2405	2430	2455	2480	2504	2529	3 5 8 2 5 7	10 13 15 10 12 15	18 20 23 17 20 22
18	2553	2577	2601	2625	2648	2672	2695	2718	2742	2765	2 5 7 2 5 7	9 12 14 9 11 14	16 19 21 16 18 21
19	2788	2810	2833	2856	2878	2900	2923	2945	2967	2989	2 4 7 2 4 6	9 11 13 8 11 13	16 18 20 15 17 19
20	3010	3032	3054	3075	3096	3118	3139	3160	3181	3201	2 4 6	8 11 13	15 17 19
21	3222	3243	3263	3284	3304	3324	3345	3365	3385	3404	2 4 6	8 10 12	14 16 18
22	3424	3444	3464	3483	3502	3522	3541	3560	3579	3598	2 4 6	8 10 12	14 15 17
23	3617	3636	3655	3674	3692	3711	3729	3747	3766	3784	2 4 6	7 9 11	13 15 17
24	3802	3820	3838	3856	3874	3892	3909	3927	3945	3962	2 4 5	7 9 11	12 14 16
25	3979	3997	4014	4031	4048	4065	4082	4099	4116	4133	2 3 5	7 9 10	12 14 15
26	4150	4166	4183	4200	4216	4232	4249	4265	4281	4298	2 3 5	7 8 10	11 13 15
27	4314	4330	4346	4362	4378	4393	4409	4425	4440	4456	2 3 5	6 8 9	11 13 14
28	4472	4487	4502	4518	4533	4548	4564	4579	4594	4609	2 3 5	6 8 9	11 12 14
29	4624	4639	4654	4669	4683	4698	4713	4728	4742	4757	1 3 4	6 7 9	10 12 13
30	4771	4786	4800	4814	4829	4843	4857	4871	4886	4900	1 3 4	6 7 9	10 11 13
31	4914	4928	4942	4955	4969	4983	4997	5011	5024	5038	1 3 4	6 7 8	10 11 12
32	5051	5065	5079	5092	5105	5119	5132	5145	5159	5172	1 3 4	5 7 8	9 11 12
33	5185	5198	5211	5224	5237	5250	5263	5276	5289	5302	1 3 4	5 6 8	9 10 12
34	5315	5328	5340	5353	5366	5378	5391	5403	5416	5428	1 3 4	5 6 8	9 10 11
35	5441	5453	5465	5478	5490	5502	5514	5527	5539	5551	1 2 4	5 6 7	9 10 11
36	5563	5575	5587	5599	5611	5623	5635	5647	5658	5670	1 2 4	5 6 7	8 10 11
37	5682	5694	5705	5717	5729	5740	5752	5763	5775	5786	1 2 3	5 6 7	8 9 10
38	5798	5809	5821	5832	5843	5855	5866	5877	5888	5899	1 2 3	5 6 7	8 9 10
39	5911	5922	5933	5944	5955	5966	5977	5988	5999	6010	1 2 3	4 5 7	8 9 10
40	6021	6031	6042	6053	6064	6075	6085	6096	6107	6117	1 2 3	4 5 6	8 9 10
41	6128	6138	6149	6160	6170	6180	6191	6201	6212	6222	1 2 3	4 5 6	7 8 9
42	6232	6243	6253	6263	6274	6284	6294	6304	6314	6325	1 2 3	4 5 6	7 8 9
43	6335	6345	6355	6365	6375	6385	6395	6405	6415	6425	1 2 3	4 5 6	7 8 9
44	6435	6444	6454	6464	6474	6484	6493	6503	6513	6522	1 2 3	4 5 6	7 8 9
45	6532	6542	6551	6561	6571	6580	6590	6599	6609	6618	1 2 3	4 5 6	7 8 9
46	6628	6637	6646	6656	6665	6675	6684	6693	6702	6712	1 2 3	4 5 6	7 7 8
47	6721	6730	6739	6749	6758	6767	6776	6785	6794	6803	1 2 3	4 5 5	6 7 8
48	6812	6821	6830	6839	6848	6857	6866	6875	6884	6893	1 2 3	4 4 5	6 7 8
49	6902	6911	6920	6928	6937	6946	6955	6964	6972	6981	1 2 3	4 4 5	6 7 8

Table B

LOGARITHMS

	0	1	2	3	4	5	6	7	8	9	1 2 3	4 5 6	7 8 9
50	6990	6998	7007	7016	7024	7033	7042	7050	7059	7067	1 2 3	3 4 5	6 7 8
51	7076	7084	7093	7101	7110	7118	7126	7135	7143	7152	1 2 3	3 4 5	6 7 8
52	7160	7168	7177	7185	7193	7202	7210	7218	7226	7235	1 2 2	3 4 5	6 7 7
53	7243	7251	7259	7267	7275	7284	7292	7300	7308	7316	1 2 2	3 4 5	6 6 7
54	7324	7332	7340	7348	7356	7364	7372	7380	7388	7396	1 2 2	3 4 5	6 6 7
55	7404	7412	7419	7427	7435	7443	7451	7459	7466	7474	1 2 2	3 4 5	5 6 7
56	7482	7490	7497	7505	7513	7520	7528	7536	7543	7551	1 2 2	3 4 5	5 6 7
57	7559	7566	7574	7582	7589	7597	7604	7612	7619	7627	1 2 2	3 4 5	5 6 7
58	7634	7642	7649	7657	7664	7672	7679	7686	7694	7701	1 1 2	3 4 4	5 6 7
59	7709	7716	7723	7731	7738	7745	7752	7760	7767	7774	1 1 2	3 4 4	5 6 7
60	7782	7789	7796	7803	7810	7818	7825	7832	7839	7846	1 1 2	3 4 4	5 6 6
61	7853	7860	7868	7875	7882	7889	7896	7903	7910	7917	1 1 2	3 4 4	5 6 6
62	7924	7931	7938	7945	7952	7959	7966	7973	7980	7987	1 1 2	3 3 4	5 6 6
63	7993	8000	8007	8014	8021	8028	8035	8041	8048	8055	1 1 2	3 3 4	5 5 6
64	8062	8069	8075	8082	8089	8096	8102	8109	8116	8122	1 1 2	3 3 4	5 5 6
65	8129	8136	8142	8149	8156	8162	8169	8176	8182	8189	1 1 2	3 3 4	5 5 6
66	8195	8202	8209	8215	8222	8228	8235	8241	8248	8254	1 1 2	3 3 4	5 5 6
67	8261	8267	8274	8280	8287	8293	8299	8306	8312	8319	1 1 2	3 3 4	5 5 6
68	8325	8331	8338	8344	8351	8357	8363	8370	8376	8382	1 1 2	3 3 4	4 5 6
69	8388	8395	8401	8407	8414	8420	8426	8432	8439	8445	1 1 2	2 3 4	4 5 6
70	8451	8457	8463	8470	8476	8482	8488	8494	8500	8506	1 1 2	2 3 4	4 5 6
71	8513	8519	8525	8531	8537	8543	8549	8555	8561	8567	1 1 2	2 3 4	4 5 5
72	8573	8579	8585	8591	8597	8603	8609	8615	8621	8627	1 1 2	2 3 4	4 5 5
73	8633	8639	8645	8651	8657	8663	8669	8675	8681	8686	1 1 2	2 3 4	4 5 5
74	8692	8698	8704	8710	8716	8722	8727	8733	8739	8745	1 1 2	2 3 4	4 5 5
75	8751	8756	8762	8768	8774	8779	8785	8791	8797	8802	1 1 2	2 3 3	4 5 5
76	8808	8814	8820	8825	8831	8837	8842	8848	8854	8859	1 1 2	2 3 3	4 5 5
77	8865	8871	8876	8882	8887	8893	8899	8904	8910	8915	1 1 2	2 3 3	4 4 5
78	8921	8927	8932	8938	8943	8949	8954	8960	8965	8971	1 1 2	2 3 3	4 4 5
79	8976	8982	8987	8993	8998	9004	9009	9015	9020	9025	1 1 2	2 3 3	4 4 5
80	9031	9036	9042	9047	9053	9058	9063	9069	9074	9079	1 1 2	2 3 3	4 4 5
81	9085	9090	9096	9101	9106	9112	9117	9122	9128	9133	1 1 2	2 3 3	4 4 5
82	9138	9143	9149	9154	9159	9165	9170	9175	9180	9186	1 1 2	2 3 3	4 4 5
83	9191	9196	9201	9206	9212	9217	9222	9227	9232	9238	1 1 2	2 3 3	4 4 5
84	9243	9248	9253	9258	9263	9269	9274	9279	9284	9289	1 1 2	2 3 3	4 4 5
85	9294	9299	9304	9309	9315	9320	9325	9330	9335	9340	1 1 2	2 3 3	4 4 5
86	9345	9350	9355	9360	9365	9370	9375	9380	9385	9390	1 1 2	2 3 3	4 4 5
87	9395	9400	9405	9410	9415	9420	9425	9430	9435	9440	0 1 1	2 2 3	3 4 4
88	9445	9450	9455	9460	9465	9469	9474	9479	9484	9489	0 1 1	2 2 3	3 4 4
89	9494	9499	9504	9509	9513	9518	9523	9528	9533	9538	0 1 1	2 2 3	3 4 4
90	9542	9547	9552	9557	9562	9566	9571	9576	9581	9586	0 1 1	2 2 3	3 4 4
91	9590	9595	9600	9605	9609	9614	9619	9624	9628	9633	0 1 1	2 2 3	3 4 4
92	9638	9643	9647	9652	9657	9661	9666	9671	9675	9680	0 1 1	2 2 3	3 4 4
93	9685	9689	9694	9699	9703	9708	9713	9717	9722	9727	0 1 1	2 2 3	3 4 4
94	9731	9736	9741	9745	9750	9754	9759	9763	9768	9773	0 1 1	2 2 3	3 4 4
95	9777	9782	9786	9791	9795	9800	9805	9809	9814	9818	0 1 1	2 2 3	3 4 4
96	9823	9827	9832	9836	9841	9845	9850	9854	9859	9863	0 1 1	2 2 3	3 4 4
97	9868	9872	9877	9881	9886	9890	9894	9899	9903	9908	0 1 1	2 2 3	3 4 4
98	9912	9917	9921	9926	9930	9934	9939	9943	9948	9952	0 1 1	2 2 3	3 4 4
99	9956	9961	9965	9969	9974	9978	9983	9987	9991	9996	0 1 1	2 2 3	3 3 4

Table C

ANTILOGARITHMS

	0	1	2	3	4	5	6	7	8	9	1 2 3	4 5 6	7 8 9
·00	1000	1002	1005	1007	1009	1012	1014	1016	1019	1021	0 0 1	1 1 1	2 2 2
·01	1023	1026	1028	1030	1033	1035	1038	1040	1042	1045	0 0 1	1 1 1	2 2 2
·02	1047	1050	1052	1054	1057	1059	1062	1064	1067	1069	0 0 1	1 1 1	2 2 2
·03	1072	1074	1076	1079	1081	1084	1086	1089	1091	1094	0 0 1	1 1 1	2 2 2
·04	1096	1099	1102	1104	1107	1109	1112	1114	1117	1119	0 1 1	1 1 2	2 2 2
·05	1122	1125	1127	1130	1132	1135	1138	1140	1143	1146	0 1 1	1 1 2	2 2 2
·06	1148	1151	1153	1156	1159	1161	1164	1167	1169	1172	0 1 1	1 1 2	2 2 2
·07	1175	1178	1180	1183	1186	1189	1191	1194	1197	1199	0 1 1	1 1 2	2 2 2
·08	1202	1205	1208	1211	1213	1216	1219	1222	1225	1227	0 1 1	1 1 2	2 2 3
·09	1230	1233	1236	1239	1242	1245	1247	1250	1253	1256	0 1 1	1 1 2	2 2 3
·10	1259	1262	1265	1268	1271	1274	1276	1279	1282	1285	0 1 1	1 1 2	2 2 3
·11	1288	1291	1294	1297	1300	1303	1306	1309	1312	1315	0 1 1	1 2 2	2 2 3
·12	1318	1321	1324	1327	1330	1334	1337	1340	1343	1346	0 1 1	1 2 2	2 2 3
·13	1349	1352	1355	1358	1361	1365	1368	1371	1374	1377	0 1 1	1 2 2	2 3 3
·14	1380	1384	1387	1390	1393	1396	1400	1403	1406	1409	0 1 1	1 2 2	2 3 3
·15	1413	1416	1419	1422	1426	1429	1432	1435	1439	1442	0 1 1	1 2 2	2 3 3
·16	1445	1449	1452	1455	1459	1462	1466	1469	1472	1476	0 1 1	1 2 2	2 3 3
·17	1479	1483	1486	1489	1493	1496	1500	1503	1507	1510	0 1 1	1 2 2	2 3 3
·18	1514	1517	1521	1524	1528	1531	1535	1538	1542	1545	0 1 1	1 2 2	2 3 3
·19	1549	1552	1556	1560	1563	1567	1570	1574	1578	1581	0 1 1	1 2 2	3 3 3
·20	1585	1589	1592	1596	1600	1603	1607	1611	1614	1618	0 1 1	1 2 2	3 3 3
·21	1622	1626	1629	1633	1637	1641	1644	1648	1652	1656	0 1 1	2 2 2	3 3 3
·22	1660	1663	1667	1671	1675	1679	1683	1687	1690	1694	0 1 1	2 2 2	3 3 3
·23	1698	1702	1706	1710	1714	1718	1722	1726	1730	1734	0 1 1	2 2 2	3 3 4
·24	1738	1742	1746	1750	1754	1758	1762	1766	1770	1774	0 1 1	2 2 2	3 3 4
·25	1778	1782	1786	1791	1795	1799	1803	1807	1811	1816	0 1 1	2 2 2	3 3 4
·26	1820	1824	1828	1832	1837	1841	1845	1849	1854	1858	0 1 1	2 2 3	3 3 4
·27	1862	1866	1871	1875	1879	1884	1888	1892	1897	1901	0 1 1	2 2 3	3 3 4
·28	1905	1910	1914	1919	1923	1928	1932	1936	1941	1945	0 1 1	2 2 3	3 4 4
·29	1950	1954	1959	1963	1968	1972	1977	1982	1986	1991	0 1 1	2 2 3	3 4 4
·30	1995	2000	2004	2009	2014	2018	2023	2028	2032	2037	0 1 1	2 2 3	3 4 4
·31	2042	2046	2051	2056	2061	2065	2070	2075	2080	2084	0 1 1	2 2 3	3 4 4
·32	2089	2094	2099	2104	2109	2113	2118	2123	2128	2133	0 1 1	2 2 3	3 4 4
·33	2138	2143	2148	2153	2158	2163	2168	2173	2178	2183	0 1 1	2 2 3	3 4 4
·34	2188	2193	2198	2203	2208	2213	2218	2223	2228	2234	1 1 2	2 3 3	4 4 5
·35	2239	2244	2249	2254	2259	2265	2270	2275	2280	2286	1 1 2	2 3 3	4 4 5
·36	2291	2296	2301	2307	2312	2317	2323	2328	2333	2339	1 1 2	2 3 3	4 4 5
·37	2344	2350	2355	2360	2366	2371	2377	2382	2388	2393	1 1 2	2 3 3	4 4 5
·38	2399	2404	2410	2415	2421	2427	2432	2438	2443	2449	1 1 2	2 3 3	4 4 5
·39	2455	2460	2466	2472	2477	2483	2489	2495	2500	2506	1 1 2	2 3 3	4 5 5
·40	2512	2518	2523	2529	2535	2541	2547	2553	2559	2564	1 1 2	2 3 4	4 5 5
·41	2570	2576	2582	2588	2594	2600	2606	2612	2618	2624	1 1 2	2 3 4	4 5 5
·42	2630	2636	2642	2649	2655	2661	2667	2673	2679	2685	1 1 2	2 3 4	4 5 6
·43	2692	2698	2704	2710	2716	2723	2729	2735	2742	2748	1 1 2	3 3 4	4 5 6
·44	2754	2761	2767	2773	2780	2786	2793	2799	2805	2812	1 1 2	3 3 4	4 5 6
·45	2818	2825	2831	2838	2844	2851	2858	2864	2871	2877	1 1 2	3 3 4	5 5 6
·46	2884	2891	2897	2904	2911	2917	2924	2931	2938	2944	1 1 2	3 3 4	5 5 6
·47	2951	2958	2965	2972	2979	2985	2992	2999	3006	3013	1 1 2	3 3 4	5 5 6
·48	3020	3027	3034	3041	3048	3055	3062	3069	3076	3083	1 1 2	3 4 4	5 6 6
·49	3090	3097	3105	3112	3119	3126	3133	3141	3148	3155	1 1 2	3 4 4	5 6 6

Table D

ANTILOGARITHMS

	0	1	2	3	4	5	6	7	8	9	1 2 3	4 5 6	7 8 9
·50	3162	3170	3177	3184	3192	3199	3206	3214	3221	3228	1 1 2	3 4 4	5 6 7
·51	3236	3243	3251	3258	3266	3273	3281	3289	3296	3304	1 2 2	3 4 5	5 6 7
·52	3311	3319	3327	3334	3342	3350	3357	3365	3373	3381	1 2 2	3 4 5	5 6 7
·53	3388	3396	3404	3412	3420	3428	3436	3443	3451	3459	1 2 2	3 4 5	6 6 7
·54	3467	3475	3483	3491	3499	3508	3516	3524	3532	3540	1 2 2	3 4 5	6 6 7
·55	3548	3556	3565	3573	3581	3589	3597	3606	3614	3622	1 2 2	3 4 5	6 7 7
·56	3631	3639	3648	3656	3664	3673	3681	3690	3698	3707	1 2 3	3 4 5	6 7 8
·57	3715	3724	3733	3741	3750	3758	3767	3776	3784	3793	1 2 3	3 4 5	6 7 8
·58	3802	3811	3819	3828	3837	3846	3855	3864	3873	3882	1 2 3	4 4 5	6 7 8
·59	3890	3899	3908	3917	3926	3936	3945	3954	3963	3972	1 2 3	4 5 5	6 7 8
·60	3981	3990	3999	4009	4018	4027	4036	4046	4055	4064	1 2 3	4 5 6	6 7 8
·61	4074	4083	4093	4102	4111	4121	4130	4140	4150	4159	1 2 3	4 5 6	7 8 9
·62	4169	4178	4188	4198	4207	4217	4227	4236	4246	4256	1 2 3	4 5 6	7 8 9
·63	4266	4276	4285	4295	4305	4315	4325	4335	4345	4355	1 2 3	4 5 6	7 8 9
·64	4365	4375	4385	4395	4406	4416	4426	4436	4446	4457	1 2 3	4 5 6	7 8 9
·65	4467	4477	4487	4498	4508	4519	4529	4539	4550	4560	1 2 3	4 5 6	7 8 9
·66	4571	4581	4592	4603	4613	4624	4634	4645	4656	4667	1 2 3	4 5 6	7 9 10
·67	4677	4688	4699	4710	4721	4732	4742	4753	4764	4775	1 2 3	4 5 7	8 9 10
·68	4786	4797	4808	4819	4831	4842	4853	4864	4875	4887	1 2 3	4 6 7	8 9 10
·69	4898	4909	4920	4932	4943	4955	4966	4977	4989	5000	1 2 3	5 6 7	8 9 10
·70	5012	5023	5035	5047	5058	5070	5082	5093	5105	5117	1 2 4	5 6 7	8 9 11
·71	5129	5140	5152	5164	5176	5188	5200	5212	5224	5236	1 2 4	5 6 7	8 10 11
·72	5248	5260	5272	5284	5297	5309	5321	5333	5346	5358	1 2 4	5 6 7	9 10 11
·73	5370	5383	5395	5408	5420	5433	5445	5458	5470	5483	1 3 4	5 6 8	9 10 11
·74	5495	5508	5521	5534	5546	5559	5572	5585	5598	5610	1 3 4	5 6 8	9 10 12
·75	5623	5636	5649	5662	5675	5689	5702	5715	5728	5741	1 3 4	5 7 8	9 10 12
·76	5754	5768	5781	5794	5808	5821	5834	5848	5861	5875	1 3 4	5 7 8	9 11 12
·77	5888	5902	5916	5929	5943	5957	5970	5984	5998	6012	1 3 4	5 7 8	10 11 12
·78	6026	6039	6053	6067	6081	6095	6109	6124	6138	6152	1 3 4	6 7 8	10 11 13
·79	6166	6180	6194	6209	6223	6237	6252	6266	6281	6295	1 3 4	6 7 9	10 11 13
·80	6310	6324	6339	6353	6368	6383	6397	6412	6427	6442	1 3 4	6 7 9	10 12 13
·81	6457	6471	6486	6501	6516	6531	6546	6561	6577	6592	2 3 5	6 8 9	11 12 14
·82	6607	6622	6637	6653	6668	6683	6699	6714	6730	6745	2 3 5	6 8 9	11 12 14
·83	6761	6776	6792	6808	6823	6839	6855	6871	6887	6902	2 3 5	6 8 9	11 13 14
·84	6918	6934	6950	6966	6982	6998	7015	7031	7047	7063	2 3 5	6 8 10	11 13 15
·85	7079	7096	7112	7129	7145	7161	7178	7194	7211	7228	2 3 5	7 8 10	12 13 15
·86	7244	7261	7278	7295	7311	7328	7345	7362	7379	7396	2 3 5	7 8 10	12 13 15
·87	7413	7430	7447	7464	7482	7499	7516	7534	7551	7568	2 3 5	7 9 10	12 14 16
·88	7586	7603	7621	7638	7656	7674	7691	7709	7727	7745	2 4 5	7 9 11	12 14 16
·89	7762	7780	7798	7816	7834	7852	7870	7889	7907	7925	2 4 5	7 9 11	13 14 16
·90	7943	7962	7980	7998	8017	8035	8054	8072	8091	8110	2 4 6	7 9 11	13 15 17
·91	8128	8147	8166	8185	8204	8222	8241	8260	8279	8299	2 4 6	8 9 11	13 15 17
·92	8318	8337	8356	8375	8395	8414	8433	8453	8472	8492	2 4 6	8 10 12	14 15 17
·93	8511	8531	8551	8570	8590	8610	8630	8650	8670	8690	2 4 6	8 10 12	14 16 18
·94	8710	8730	8750	8770	8790	8810	8831	8851	8872	8892	2 4 6	8 10 12	14 16 18
·95	8913	8933	8954	8974	8995	9016	9036	9057	9078	9099	2 4 6	8 10 12	15 17 19
·96	9120	9141	9162	9183	9204	9226	9247	9268	9290	9311	2 4 6	8 11 13	15 17 19
·97	9333	9354	9376	9397	9419	9441	9462	9484	9506	9528	2 4 7	9 11 13	15 17 20
·98	9550	9572	9594	9616	9638	9661	9683	9705	9727	9750	2 4 7	9 11 13	16 18 20
·99	9772	9795	9817	9840	9863	9886	9908	9931	9954	9977	2 5 7	9 11 14	16 18 20

INDEX

A
accuracy 19-32
  spurious 30-1
addition 28, 61-5
administration 11
algebra 70
Annual Abstract of Statistics 11, 14
approximation 19-32, 43
arithmetic mean 129-37
arithmetic progression 72-3
averages 129-50
  arithmetic mean 129-37
  geometric mean 148
  harmonic mean 148
  median 137-41
  mode 144-8

B
Bank of England Quarterly Bulletin 14
Bank Review 14
bar charts 99-103
  component 102
  compound 101
  horizontal 99-100
  percentage 103
  simple 99-100
bias 22-6, 30-1
  interviewer 39-40
binary 61
BODMAS 62
break-even charts 120-1
British Business 13
BUSINESS STATISTICS OFFICE 12
buying 16

C
calculators 6, 75-6
cartograms 109-10
Census of Distribution 14
Census of Population 8, 14, 15, 43
Central Limit Theorem 46, 181
CENTRAL STATISTICAL OFFICE 12-16

characteristic 76-9
class interval 85-7
classification 85-7
climate 13
coding 35
coefficient of correlation 204-9
coefficient of variation 170
confidence limits 183-4
continuous variables 85
correlation 198
  Pearson's 204-6
  product moment 204-6
  rank 206-7
  spurious 208-9
  tables 203-4

D
data 1-8, 82
  primary 6-8, 82
  secondary 6-8, 82
decimals 60-1, 65-7
decisions 75, 173-87
degrees of tolerance 20-1
dependent variable 111-13
deviation 151-72
discounted cash flow 74-5
discounted rate of return 75
discrete variable 85
dispersion 151-72
distribution 13
  of sampling means 175
distributions 151-7, 160
  bi-model 153
  J-shaped 154
  normal 152, 161-2, 175-9
  rectangular 153
  skewed 152-3
division 29, 62, 64-5, 79

E
Economic Progress Report 14
Economic Trends 13
education 15

Employment Gazette 14
error 21-6
  absolute 24-5
  biased 25
  compensating 26
  cumulative 25
  relative 25
  standard 179-84
  systematic 25
  unbiased 26
estimation 173
extrapolation 113, 211

F

Facts from Your Figures 13
Family Expenditure Survey 8, 14, 56, 194-5
finance 13, 16
financial mathematics 72-5
Financial Statistics 13, 72-5
forecasting 211-28
fractions 63-5
frequency curves 96-8
frequency distributions 87-9, 132-6
frequency polygons 94-6

G

Gantt charts 119-20
geometric mean 148
geometric progression 72-8
Government statistics 11-14
graphs 110-19, 126-8
  semi-log 116-18
  straight-line 118-19
Guide to Official Statistics 12
Guinness Book of Records 13

H

harmonic mean 148
HER MAJESTY'S STATIONERY OFFICE 12
histograms 90-4, 166
hypothesis tests 181-3

I

imperial system 66-7
independent variable 111-13
index numbers 188-97

index numbers (*contd*)
  chain-based 192
  Laspeyre 192
  Paasche 192
  weighted 190-2
Index of Retail Prices 14, 188, 190, 194-5
indexes 14
industry 10-11, 14-15, 43
inflation 3-4
information 1-10, 10-18, 19-32, 33-42, 82
  in business 10-11
  macro-statistical 10
  micro-statistical 10
intercept 222-4
interest 72-3
  compound 72-3
  simple 72
interpolation 211
interquartile range 156-9
interval scale 71
interviewing 37-40
  formal 38
  informal 38-9
  interviewer bias 39-40

L

labour 13
law of the inertia of large numbers 46
law of statistical regularity 45-6
least squares 222-6
libraries 15
line charts 91
linear trends 221-6
  least-squares 223-6
  semi-averages 221-2
  three-point 221-2
logarithms 76-81, 231-4
Lorenz Curves 123-5

M

Management Today 14
mantissa 77-8
manufacturing industry 10-11, 14-15, 43
map charts 109-10
marketing 11, 16, 43
mathematics 60-81

mean 129-37
mean deviation 163
measurement 70-1
median 137-41
metric system 66
modal class 146-8
mode 144-8
Monthly Digest of Statistics 13, 15
multiplication 29, 62, 64-5, 78-9
Municipal Year Book 13

N

national economy 10, 13-14
National Income and Expenditure 'Blue Book' 13, 14
net present value 74-6
nominal scale 70
normal curve 152, 161-2, 172-9

O

observation 35-7
  mechanical 37
  participant 36
  systematic 36
OFFICE OF POPULATION AND CENSUS AND SURVEYS 12, 15
official statistics 10-18
  problems 15-16
  profiting from 16-18
ogive 139-41, 157-9
ordinal scale 70-1
overseas statistics 13

P

panels 58-9
percentages 33, 67
perception 125-9
personnel 11, 16
pictograms 107-9
  comparative 109
pie charts 104-7
  comparative 106-7
population 13, 20, 26-7
powers 68-9, 79
present value 74-6
presentation 82-128
prices 3-4, 13
primary data 6-8, 82
probability 173-5, 178-9

production 11, 13
Profit from Facts 13, 16
proportions 69
public finance 13
public services 13

Q

quality control 184-7
quartiles 141-4
questionnaires 40-1
  design 40-1
  postal 40

R

random numbers 49
random samples 49-50
range 155-6
ratios 69
regional statistics 14
REGISTRAR-GENERAL 12
regression 209, 221-6
reports 35, 89-90
roots 68-9, 79
rounding 22-4

S

sampling 43-59
  bias 48-9
  cluster samples 54-5
  design 47-8
  error 46
  frame 47-8
  interpenetrating sampling 57-8
  master samples 58
  multi-phase 57
  multi-stage 55-7
  panels 58-9
  population 43, 47
  quota 52-4
  random route 51
  replicated 57-8
  simple random 49-50
  size 47
  stratified 51-2
  systematic 50
  units 47-8
sampling distribution of the means 175
scatter diagrams 200-3
seasonal variations 215-19
secondary data 7-8

semi-averages 220-2
significance tests 181-3
significant figures 24
skew 152-3
slope 222-4
social statistics 13
Social Trends 14
square root 68-9, 235-8
standard deviation 159-72, 180-1
standard error 179-84
Standard Industrial Classification 14-15
standard normal distribution 175-8
Statesman's Year-Book 13
statistics
  abuse of 4-5
  definition of 1
  descriptive 6, 44-5
  inductive 6, 45
  secondary 8-9, 35
  use of 5-6
Statistical News 13
strata charts 110-11
subtraction 28-9, 62, 63-4
surveys 33-5
  design of 34
  government 34
  market research 34
  objectives of 33-4
  pilot 34-5
  systematic 33-4
symbols of mathematics 80

T

tables, statistical 35, 83-5
time series 212-20
  irregular fluctuations 220
  moving averages 213-15
  residual fluctuations 220
  seasonal variations 215-19
trends 211-28
truncation 24
type I and II error 183

U

U.K. GOVERNMENT STATISTICAL SERVICE 12
unemployment 8

V

variables 111-13
  dependent 111-13
  independent 111-13
variance 169
  coefficient of variation 170
vital statistics 13

W

Whitaker's Almanack 13

Z

Z-chart 122-3, 230
Z-values 177-9